About The Book

DNS is a foundational element of network communications. It's also the starting point for countless cyberattacks. Threat actors abuse DNS to install malware, exfiltrate data, and perpetrate malware threats. Cyber threats that leverage DNS are widespread, sophisticated, and rapidly evolving. DNS is used by over 90 percent of malware and in an ever-growing range of pernicious attacks.

However, despite its vulnerabilities, DNS can unlock a hidden world of security capabilities that can help protect today's highly distributed and cloud-integrated networks. ***The Hidden Potential of DNS in Security*** reveals how attackers exploit DNS and how cybersecurity professionals can proactively use DNS to turn the tables and mitigate those threats.

Knowing how to leverage the protective capabilities of DNS can give you an unprecedented head start in stopping today's advanced cyberthreats. This book gives you that knowledge.

Written specifically for security practitioners, and including real-world case studies, this book offers a thorough yet easy-to-digest understanding of today's most urgent and potentially damaging DNS-based cyberthreats, how to mitigate them, and how to leverage your DNS infrastructure to further your security mission. In it, you will discover:

- Why DNS is inherently vulnerable and why knowledge of DNS is now crucial for security teams
- How malware uses DNS to avoid detection and communicate with command-and-control (C2) infrastructure
- How threat actors leverage DNS in executing a broad array of attacks involving look-alike domains, domain generation algorithms, DNS tunneling, data exfiltration, and cache poisoning
- What DNSSEC is (and is not) and how it works
- How recently emerging encrypted DNS standards can impact security controls, along with the security advantages they can provide
- How DNS can be leveraged in Zero Trust architectures
- How you can improve your security posture using the DNS infrastructure you already have

The Hidden Potential of DNS in Security

Joshua M. Kuo and Ross Gibson, J.D.

First Infoblox paperback edition July 2023

Infoblox publishes its books in a variety of electronic and print formats. Some content that appears in print may not be available in electronic books, and vice versa.

ISBN: 978-1-960940-01-8

Table of Contents

Foreword

In 1996 I was working DNS support for a large ISP when I got a call from Verisign. This was two years before the Internet Corporation for Assigned Names and Numbers (ICANN) was established, and Verisign was the authoritative registry for all generic top-level domains. The nice person from Verisign on the other end of the line was trying to reach someone in our organization because our domain had expired 30 days ago, and it was about to be deactivated. They had been trying to get someone, anyone, to respond to them to renew the domain.

This was a serious problem. If our domain expired, it would have disrupted our routing. It would have prevented us, and our clients, from sending email. Our UUCP feed would be interrupted, and our few hosting clients would not have been able to log into their servers. In short, our domain expiring would have shut us down. I escalated the call to my manager, who pulled out his personal credit card and paid the $100 to keep the domain live for a year. From there, we set up a meeting with the teams responsible for managing our DNS servers, our marketing team and leadership to find out who was going to be responsible for renewing the company's domains going forward. The plan was put in place, and we never had another lapse. I don't know if my manager ever got reimbursed.

This was an early lesson for me on the importance of DNS security—a lesson that, unfortunately, many organizations still need to learn. That is why this book from Joshua M. Kuo and Ross Gibson, J.D. is so important. DNS security is one of those issues that is too often ignored, ill-defined or misunderstood by many organizations.

In fact, many organizations don't know where to start with DNS security because DNS has a lot of areas of vulnerability and exposure. This is why I love that this book covers a wide swath of DNS security areas from response policy zones (RPZs) to DNS being used for exfiltration.

The way the authors laid out their book makes it easy to follow along, starting with all the ways that DNS is insecure and why it is so important for defenders to understand DNS basics (which, sadly, too many don't). This setup leads nicely into how malware uses DNS, including communicating with C2 servers. Then, the book discusses look-alike domains that threat actors use in phishing and other campaigns and how domain generation algorithms (DGA) factor into the picture. Readers are then taken on the journey of DNS tunneling and data exfiltration and, more importantly, how the limits of DNS can help defenders facilitate detection.

Later chapters provide a valuable overview of DNS cache poisoning, an attack type that has been around for decades and still catches some people off guard. The book also delves into DNSSEC and how it can help prevent not just cache poisoning but other types of attacks as well. Encrypted DNS, a newcomer that's not widely understood, is also covered, along with what security teams need to know about how it can impact detection and policy enforcement.

The Hidden Potential of DNS in Security also looks at different DNS clients and how they can be susceptible to attack, including the (still) vulnerable hosts.txt file. The book then tackles the term "DNS hijacking," what it actually means, and how to defend against it. DNS hijacking is a confusing term to a lot of people, so it is good to see it so clearly explained. The authors end by summarizing some next steps for security professionals, taking DNS and network security to the next level with D-NAP.

One of the most important aspects of the book is that it doesn't just lay out the threats; it also provides practical solutions to defend against those threats. The authors have been doing this for a while and it shows with many strategies that can be readily implemented.

I am glad you have picked up this book. It will help your security teams and organizations add another layer of defense against the ever-expanding threat landscape.

— *Allan Liska, Intelligence Analyst at Recorded Future and co-author of* DNS Security: Defending the Domain Name System

Acknowledgements – Joshua M. Kuo

This book was more than 2 years in the making, and many people helped to make it a reality. I would like to first thank Brian “Cheebert” Chee, who opened the door to the world of technology for me and helped me forge many life-altering connections. Curtis Franklin Jr., thank you for believing in a young man who wanted to write all those years ago. Jim Martin, thank you for your honest feedback, especially for your insights on the organizational structure of ICANN and IANA.

A special thank you goes to Cricket Liu, for always being willing to share your wisdom and friendship. Cricket connected me to my co-author Ross, and with many key individuals. Without you, Cricket, this book literally would not have been produced. To my co-author Ross Gibson, I am grateful that we got to work on this project together, your attention to detail and insight added tremendous value. I had fun working on this together with you; I hope we get to do it again in the near future.

I would like to recognize Piotr Głaska for the original framework with case studies, Bryan Norman for helping with viable book titles, and Kate McDonald for discussions and introducing me to DNS many years ago. Thanks to all the reviewers who spent valuable time and effort to read our drafts and manuscripts: Anthony Ciarochi, Renyk de’Vandre, Seth Hammerman, Steve Gibson, Allan Liska, Steve Makousky, and Andreas Taudte.

Special thanks to the Infoblox team for making this project possible: Karen Bissani, Steven Carr, Geoff Horne, Tan Huynh, and Sandy Johnson. Thank you, Laurie Robison, for making yourself available to those early morning meetings before sunrise. Thanks to the editors, Sean Kirk and Pat Tompkins, for your patience and diligence.

Finally, I would like to thank my family: Cynthia, Ian, Isaac, and Illiana. There were many days and nights that I devoted myself to this project instead of spending time with you; thank you for your support.

Acknowledgements – Ross Gibson, J.D.

What an interesting journey this project has been. I wish I were able to thank everyone who has helped me get to this point, but the list would be longer than the book, so sincere apologies to anyone I left out.

First, I need to thank two gentlemen without whom this book would never have come to life. To my co-author, Josh, the drive and commitment you displayed to keep the project moving forward was incredible, and kept me inspired throughout the process. In addition to that, your talents as a writer and educator shined through to make the entire book the fantastic resource it has turned out to be. Tremendous thanks must also go to Cricket Liu, for bringing the two of us together, giving feedback on drafts, sharing invaluable insight into the process of writing a book, and providing guidance on how to get the job done. I will be forever grateful for his friendship and support.

Many thanks are due to the other members of the team who made this possible, our editors Sean Kirk and Pat Tompkins, marketing team Karen Bissani and Laurie Robison, and graphics masters Tan Huynh and Santosh Kamble. Special thanks to Piotr Głaska for his foundational research and work that formed the basis for much of the information presented here. To all the reviewers who took time to read through the manuscript and provide valuable feedback, we are deeply indebted to you.

To my family, Jenn, Pierce, and Evan (and our dog, Aspen), my role in this project could not have happened without your love and support. To Jenn especially, thank you for the sacrifices you made to support me in this project and several others throughout our years together. I love all three of you. To my parents, Hank and Susan Gibson, and my sister, Kate, I wouldn't be who and where I am today without you; thank you for everything.

To my long-time DNS brother, John Steele, thanks for providing initial reviews and always being available as a sounding board for ideas. I'm honored to be able to call you a friend and teammate stretching as far back as the days of EDDI.

Thank you to the Infoblox leadership team, especially to Geoff Horne and Victor Danevich for enthusiastically supporting my involvement in the project. Having the opportunity to work directly with you both has been an absolute pleasure and a learning experience of incredible value. To my global SME teammates, I'm humbled to be able to work alongside each of you. To my former account team partners, Jeff Cummings, Jamie Jaseph, Lesley Ransom, Al Zanger, and Sarah Walton, thank you for being such great teammates and travel companions throughout all our journeys together. To my SE/SA managers throughout the years, Don Smith, Alejandro Aguila, Allen McNaughton, Mark Reyero, Chris "DC" Elmore, and Ken Albanese, thank you for your support and guidance. Thanks also to Sandy Johnson for his constant support of the project. Sincere thanks go out to the customers who kept providing problems to solve and entrusting me to help provide a solution.

To my original DNS teacher, Pete Siebentritt, first DNS partner, Michael Simoni, and years long partner-in-crime Jeff Tejnecky along with the rest of my Capital One DNS colleagues, Asim Mubashir, Tom Burgoon, Tim Coppedge, John Cotney, Jason Davis, Lynn Hilton, Charles Lee, John Oquendo, and Mike Good among others, thank you for your support over the years helping to hone my DNS knowledge and skills. Thanks also to Steve Donohue for giving me the opportunity to dive into the DNS pool and for the management support of other leaders over the years like Ted Polito, Vince Gutosky, Jerome Pudwill, Mark Colangelo, Loren Morgan, Eric Bane, Steven Gray, and Leon Li. I also greatly appreciate the support over the years of several colleagues like Brad Jones, Bob Relyea, Dan Worthley, Mike Gebhardt, Glen Wilson, Don Munyak, Jay Strickland, Bill Lopez, and Jeanmarie Bright. Thanks to the international teams that were so welcoming to me over the years, especially Jon Allen, Jake Lee, James Watson, Andy Barraclough, Dan Anderson, Bo Gorham, Beau Hartman, Ben Marples, Mark Taylor, Neil Jones, and Ammo Gill. Special mention must be made of Phil Mason, who provided an opportunity to a recent college grad interested in networking. Thanks also to my friend and colleague, Jake Kouns for his support and encouragement in many areas and forms inside and outside of work over the years.

And to you, the reader, I want to thank you for taking the time to read what many people have worked hard to produce. I sincerely hope you find many valuable pieces of knowledge herein.

CHAPTER 1

Introduction

How to Read This Book

Who This Book Is For

This vendor-agnostic book is written for security professionals who want to understand the risks and rewards of the DNS (Domain Name System) in a security context. We understand that DNS is not something most security practitioners spend a great deal of time studying, but it is nonetheless a critical component that security professionals must understand. Thus, this book focuses on the use of DNS by a security practitioner rather than how a system administrator would secure a DNS infrastructure. If you are a security professional trying to determine how to best leverage your DNS infrastructure to further your security mission, this book is for you.

We assume you, the reader, are familiar with basic networking concepts and terms such as ICMP, UDP, TCP, IP addressing, and ports. However, we do not assume you have any understanding of DNS beyond the fact that it maps domain names to IP addresses and vice versa.

Who This Book Is Not For

If you have picked up this book hoping to learn how the DNS protocol works, or how to set up a DNS server, this is not the right book[1] for you. If you want to learn about DNSSEC, this book will give you a very high-level overview of what problem is solved by DNSSEC but not nearly enough details for you to learn how it works or how to implement it. If you want to learn how to protect your DNS infrastructure from cyber attacks, we only touch lightly on that topic. For more detail, you should seek out books or resources specific to the flavor of DNS servers you seek to protect.

Life as a Security Professional

You are a security professional. Your title could be anything from "Security Engineer" to "Chief Information Security Officer," but what you do day to day is similar: defend your organization from threat actors. And here is something you probably don't want to admit, at least not out loud: The threat actors have a lead. Your adversaries are not smarter or better, but attackers outnumber defenders.

Another disadvantage you face is that you need to cover all potential entry points, but the attackers only need to find one opening. You have alert fatigue from your intrusion detection system (IDS), next-gen firewall, and various other probes scattered throughout your network and systems. It's impossible to adequately address all these notifications, much less to know which ones to act on. "If I could just have 10 more people on my team," you may think, but you also know in your

---

1 The right book for you is *DNS and BIND*, authored by our friend and colleague Cricket Liu.

heart that it's just a matter of time before the attackers catch up. Then you'll need 20 or 30 more people. Whatever your plan may be, threat actors are evolving their techniques, and so should you. Think of this book as a powerful resource (on an often-overlooked subject) that you can add to your security tool kit.

What Types of DNS Exploits Are Covered in This Book

This book explores the ways in which nefarious actors exploit various insecurities in DNS. It is admittedly a sprawling topic. Many security experts have their own opinions about how to talk about DNS-based threats. For the purposes of our discussion and to help security practitioners better understand how these threats work, we have organized them into five main types. The first two deal primarily with the underlying DNS infrastructure. Because there are many other resources addressing risks associated with these two types, we do not cover them here. Instead, our focus is on the other three—the ones that (ab)use the DNS protocol itself.

Type 1: Service Impediment. This category of DNS attacks focuses on slowing down or taking down the DNS service. This includes denial of service (DoS), amplification and reflection, and phantom domains. There are other resources covering this category, so we do not address them in this book.

Type 2: Implementation Vulnerabilities. This category of DNS attacks is usually specific to the type of DNS software and version that is running the service. Such attacks include remote code execution, privilege escalation, and vulnerabilities such as CVE-2020-1350 (Microsoft SIGRed) and CVE-2021-25216 (BIND buffer overflow). We do not cover these in this book.

Type 3: DNS as a Transport. This category of attacks does not aim to stop or slow down the DNS service. Rather, bad actors use DNS messages as carriers, disguising malicious payloads as innocent DNS messages and sending them in or out of the premises. We describe these techniques in detail in Chapters 5 and 6, "DNS Tunneling" and "Data Exfiltration."

Type 4: Redirection (or Misdirection). There are many names for this category, such as DNS doctoring, DNS spoofing, and DNS hijacking. Unfortunately, none of these terms are well defined or standardized in the industry. The attack tactic is simple: You think you are visiting X, but really you are visiting Y. We describe these techniques in Chapters 3, 7, and 10, "Look-Alike Domains," "Cache Poisoning and DNSSEC," and "Domain Hijacking." The technique itself is not necessarily malicious in nature, in fact, as we see in Chapter 3, one of the main technologies to combat DNS insecurity, Response Policy Zone (RPZ), uses redirection to divert users away from malicious names.

Type 5: Data Privacy. The DNS protocol itself has fundamental flaws that result in the lack of data privacy. Attacks that leverage this pathway have many components, including DNS cache snooping (performed against DNS recursive servers), zone transfer and information disclosure (performed against DNS authoritative servers), and privacy leakage through DNS queries themselves. Bad actors can use these techniques to probe DNS servers and learn about the intended victims. In some cases, this type of reconnaissance can even be performed off-path. We do not cover this threat type comprehensively in this book. One aspect of it we do cover, however, is privacy during transport, included in Chapter 8, "Encrypted DNS."

How This Book Is Organized

When writing this book, we recognized that today's security professionals don't have a lot of time, so we opted to focus on basic information instead of the level of detail that would lead to a 600-page book. We've organized the information in a logical flow. You will benefit most by reading each chapter in sequence because some of the concepts build on top of ones mentioned previously.

Because DNS itself can be complex and quite nuanced, we try to assume as little prior knowledge about DNS as possible. Instead of making the first half of this book about "how DNS works," we instead introduce just enough DNS knowledge to understand each particular security topic. We spell out each abbreviation or acronym that we use, except for the most common ones, such as TCP, HTTP, and ISP.

In Chapter 1 (this chapter), we provide a brief overview of why DNS is insecure and why DNS knowledge is relevant to today's security practitioners.

In Chapter 2, "DNS and Malware," we discuss how malware uses DNS to avoid detection and communicate with command-and-control (C2) servers.

In Chapter 3, "Look-Alike Domains," we examine different techniques that threat actors employ to hide their tracks and make detection difficult. We also learn some technologies and techniques that can help us combat the ever-growing number of malicious domain names.

In Chapter 4, "Domain Generation Algorithms," we look at techniques that threat actors use to make detecting malicious activities even more difficult, which may require defenders to switch to a different approach.

In Chapter 5, "DNS Tunneling," we discuss how malware uses DNS to send and receive data over DNS. We also look at how the DNS protocol limits the ways that threat actors build DNS tunnels, paving the way for more accurate detection.

In Chapter 6, "Data Exfiltration," we describe how threat actors use DNS to exfiltrate (and infiltrate) information and what characteristics to look for to help detect and stop the exfiltration.

In Chapter 7, "Cache Poisoning and DNSSEC," we turn our attention from DNS messages to DNS servers. We focus on the particularly damaging cache poisoning threat against recursive resolvers and the cure that DNSSEC can potentially provide.

In Chapter 8, "Encrypted DNS," we discuss the recent emergence of user data privacy tools in DNS and its two most widely used forms: DNS over TLS (DoT) and DNS over HTTPS (DoH). We discuss the ramifications of these new technologies for security professionals.

In Chapter 9, "DNS Attacks Against Clients," we profile threats that target DNS clients, including HOSTS file tampering, cache manipulation, and redirection to unauthorized DNS servers.

In Chapter 10, "Domain Hijacking," we clarify this ill-defined term, which is often used to refer to several different types of attacks, including ones against domain registrars and cloud providers.

In Chapter 11, "DNS and Zero Trust Architecture," we provide an overview of the ways that DNS-based security strategies outlined in this book can help you achieve Zero Trust principles.

In Chapter 12, we summarize what we have learned and provide a list of practical DNS-related steps you can take to enhance your overall security posture.

In the "This Book at a Glance" section, we give you a quick reference guide to the material covered in more detail in each of the chapters.

Why DNS and Why Is It Insecure?

DNS is ubiquitous. It is generally the first network protocol that devices invoke in network interactions. It also happens to be inherently insecure. When DNS was first created in the 1980s, its primary goal was to make it easier for users to connect to resources over a network without having to remember long strings of numbers. It was designed as an open protocol that would be relatively simple to manage. Security safeguards were not built into DNS. As with social media platforms that have come afterwards, the original creators of DNS never imagined how it would one day be abused.

Since the inception of DNS, cyber attacks that leverage the protocol have mushroomed. In 2017, the SANS Data Protection Survey[2] showed that nearly 30 percent of data breaches occurred over DNS. In 2019, Accenture reported that 91 percent of malware uses DNS. In 2020, the U.S. National Security Agency (NSA) stated that a protective DNS (PDNS) solution could help mitigate up to 92 percent of malware attacks.[3] The NSA and CISA describe PDNS as "different from earlier security-related changes to DNS in that it is envisioned as a security service—not a protocol—that analyzes DNS queries and takes action to mitigate threats, leveraging the existing DNS protocol and architecture."[4] In Chapters 3 and 6, we discuss some of the foundational technologies that are at the heart of most PDNS solutions.

DNS was not always such a prime target. For many years, HTTP/S was the leading attack vector. A great deal of energy went in to shoring up client browsers, web servers, and the protocol itself. These days, attackers know that HTTP/S is under strict scrutiny; even encrypted traffic is likely decrypted for inspection. Products such as web application firewalls (WAFs) are taking apart HTTP messages for deep inspection. Some people used to joke that HTTP/S was the "great firewall bypass," but it is no longer true. Today, if a web page tries to infect the user's web browser with a trojan, there are many safeguards along the way to stop it.

For most organizations, the same level of safeguards has not been deployed for DNS, at least, not yet. DNS is the second most widely used network protocol, after HTTP/S, but DNS is one of those foundational technologies that tends to fly under the radar of most cyber security teams. Many attackers know that DNS is not being watched as closely as HTTP/S and that consequently, sending and receiving malicious payloads over DNS is less likely to be detected or blocked.

In truth, DNS is so widely abused today that we've devoted this entire book to the many ways attackers leverage the protocol for malicious purposes. Our goal is to provide an overview of the security risks inherent in DNS, along with areas you may wish to research more fully to take advantage of the hidden potential in DNS for shoring up your defenses.

A Word About Case Studies in This Book

It is not our intention to list all the malware that (ab)uses DNS. That list could be longer than this book itself! We have selected examples of specific malware and security incidents that best illustrate how a given type of DNS attack unfolds. There was no shortage of examples for us to choose from. The (ab)use of DNS

---

2 SANS (2017) Sensitive Data at Risk Everywhere: The SANS 2017 Data Protection Survey [White paper]

3 https://www.nextgov.com/cybersecurity/2020/06/nsa-piloting-secure-domain-name-system-service-defense-contractors/166248/

4 https://media.defense.gov/2021/Mar/03/2002593055/-1/-1/0/CSI_Selecting-Protective-DNS_UOO11765221.PDF

is so prevalent among malware today, it is safe to assume any malware you encounter is leveraging DNS in one way or another.

A DNS Primer for Security Professionals

Before we dive headfirst into the subject of DNS-based threats and tactics to thwart them, we need to define some foundational terminology within the DNS context and present some information about how DNS functions. The improper use of terminology can lead to confusion at best and an operational outage at worst, so it is wise to be precise. We hope that after reading this book you will stand out as a beacon of proper DNS terminology. We're also going to define some terms for concepts that may lack generally accepted meanings today because, well, someone has to at some point!

DNS, DNS, DNS Everywhere!

If you step back and consider the role DNS plays in enabling technology that drives modern society, it is easy to forget that DNS had humble beginnings. It began as a way for people to more easily find computers on the ever-expanding ARPANET by entering a name, rather than having to remember IP addresses. At its most basic level, DNS provides mapping of domain names to IP addresses (much like the Contacts app on your smartphone). The process of determining the IP address mapped to a given name is called *resolution*.

When Paul Mockapetris formally documented the DNS design in 1983, sharing and trust were part of network culture. DNS security was on no one's radar back then, but it is now. DNS-based threats lurk around every corner. Security needs to be a primary consideration for any modern system, and DNS is a powerful tool when employed in a security role.

DNS is a foundational technology—once a network is built, things don't really start happening on that network unless DNS is available. Most network communications start with a DNS query: the initial request for the IP address that is returned for a given domain name. DNS queries are sent between DNS servers, which translate the domain names to their corresponding IP addresses. Keeping those query resolutions humming along briskly requires legions of DNS servers. The pervasiveness of DNS, coupled with its unique position within networks (close to endpoints), are among the reasons why DNS plays such a central role in a modern cyber-security ecosystem.

Unfortunately, the ubiquity of DNS, coupled with its historic lack of security controls, are what have made it such a prevalent attack vector today. If you fail to secure your DNS infrastructure, threat actors will use that infrastructure against you (and others).

DNS infrastructure can be leveraged to give a security team data that provides valuable contextual information when investigating potential security incidents.[5] While that data can be incredibly valuable on its own, it can be combined with Dynamic Host Configuration Protocol (DHCP) and IP address management (IPAM) data to create a gold mine for security operations and incident response teams. This book supplies guidance about how you can gather, analyze, and use this information to your greatest advantage. When used properly, DNS can go from your weakest link to your first line of defense.

DNS Is Complex and Ever Evolving

As anyone who studies DNS will tell you, DNS terminology[6] is confusing. Furthermore, it is still evolving. Take one of the first DNS security features, transaction signature (TSIG), for example. It was first released in the year 2000 and went into widespread use. However, 20 years later when everyone thought this feature was set in stone, its specification was rendered obsolete by a new set of standards.[7] DNS is sometimes impossible to keep up[8] with, even for DNS experts.

Fret not! We know you are here to learn about security, not DNS. If you want to learn about DNS, there are many excellent books and resources (such as the seminal *DNS and BIND*, by our esteemed colleague and friend Cricket Liu). In this section, we try to cover the minimum amount of DNS knowledge necessary for us to proceed with our security topics. Later in the book, whenever additional background information about DNS is required to understand a particular concept or type of attack, we provide those details within that topic.

URL or Domain Name?

First, let's talk about the difference between a web address, or Uniform Resource Locator (URL), and a domain name. Due to the prevalence of the HTTP/S protocol, people sometimes mix up these terms. Figure 1 shows the anatomy of a URL. The domain name is only one portion of the URL. Additional URL components, such as the path, have nothing to do with DNS. In the figure, `www.example.com.` is the domain name portion of the URL.

---

5 That is, assuming you retain an appropriate amount of DNS logging data. In very busy environments, this can be vast amounts of data, so the costs associated with that retention should be considered carefully.

6 The latest DNS terminology is described in RFC 8499 https://datatracker.ietf.org/doc/html/rfc8499.

7 TSIG was originally specified in RFC 2845, later respecified in RFC 8945.

8 The DNS Camel project aims to be a central hub for all the most up-to-date DNS RFC's and specifications. https://powerdns.org/dns-camel/

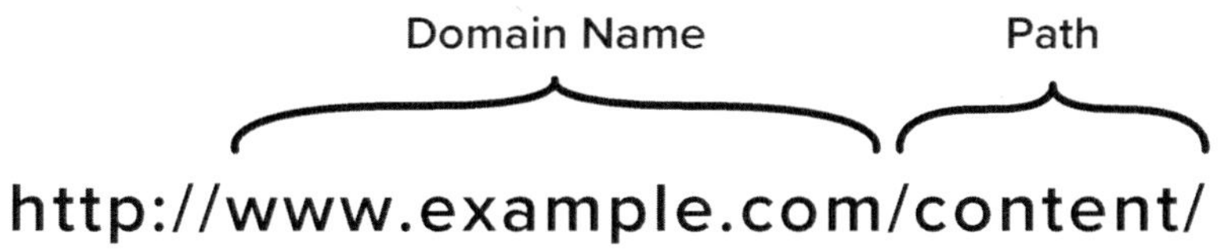

Figure 1: Domain name within URL

Anatomy of a Domain Name

Figure 2 breaks down the anatomy of a domain name, which is also commonly referred to as a DNS name, DNS entry, or DNS record. The proper format of the domain name is to have a trailing dot that represents the DNS root, although almost all DNS implementations can interpret a domain name without the trailing dot. This is similar to how browsers now automatically add http:// or https:// to the beginning of the URL. Each part of the domain name that is divided by the dot (.) character is known as a label. The first label in the name (i.e., farthest left) is known as the host name, and that part of the domain name is sometimes called the host portion of the name. The last label, the one closest to the root (the trailing dot), is also known as a top-level domain (TLD).

Why is the trailing dot important? When every label of the domain name is written out and the domain ends in a trailing dot, we call it a fully qualified domain name, or FQDN. The trailing dot directly ties the domain name to the root and is essential to creating the FDQN. An FQDN is the complete name and refers to a unique address on the Internet. If you want to be certain that you are accessing the right resource, you need a trailing dot at the end of the name. Note however, that final dot is often left off by non-DNS professionals (most of us). If the trailing dot is not used, it is possible that certain software (such as most operating systems) will attempt to add suffix information during name resolution, making your domain name potentially ambiguous. For example, when you attempt to look up `xyz.example.com`, without the trailing dot, the underlying system or software may assume: "Oh, you must mean `xyz.example.com.company.local.` instead. Let me look that up first."

It is best to use the trailing dot, when possible, to define the end of the domain name to avoid any potential confusion or unintended resolution to a destination the user is not seeking. The trailing dot notation is also how you will see FQDNs written in most DNS log files and packet captures in your security investigations, so it is good to understand its meaning. In the spirit of being precise, we will attempt to use this notation style in most places throughout the book.

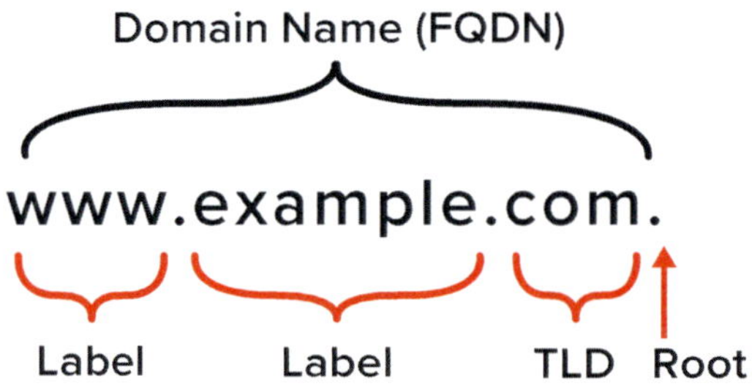

Figure 2: Domain Name Components

Components of the Domain Name System

To understand DNS-based cyber threats, it's helpful to have a general understanding of the components of DNS. As we'll show you in subsequent chapters, DNS attacks exploit one or more of these components, which include:

- Authoritative servers
- Recursive resolvers
- Resource records
- Namespaces
- Stub resolvers

Authoritative Servers

When most people think of a DNS server, they usually think of it as one giant, monolithic beast that does everything. In reality, it's a multi-headed beast. As Figure 3 illustrates, on the right side, we have *authoritative servers*, which store definitive answers to DNS queries for a given group of domains and subdomains (like a database).

Recursive Resolvers

In the middle of the diagram, we have *recursive resolvers*, which traverse the Internet to hunt down the answers to queries. In many cases, these recursive resolvers have the answers stored in cache. When they don't, they will send a query on to other recursive resolvers or to a series of authoritative servers until the answer is found, at which point the recursive resolver returns the complete answer to the client.

One DNS server can play multiple roles. However, for clarity's sake in describing the process, we have segregated each role to its own server in the diagram. For example, a DNS server can be authoritative for some domains while also providing recursive resolution for other domains, which is commonly seen inside corporate networks.

Resource Records

An individual pairing of DNS data points, such as the name `www.example.com.` and its IPv4 address, 93.184.215.34, is known as a resource record. People frequently refer to resource records as subdomains, which is incorrect. Subdomains are containers that resource records reside in, not resource records themselves. For example, in the domain name `www.example.com.`, `example` is a subdomain of `com.`, and www is a resource record within that subdomain.

Namespaces

Within an organization, all the authoritative domains under its control and the DNS resource records within them are collectively referred to as its *namespace*. For example, a corporation could operate `example.com.`, `example.org.`, and `example.local.`, and its namespace would consist of all three domains.

Stub Resolvers

As for the DNS client on the left in Figure 3, we often refer to it as the *stub resolver*. The DNS stub resolver (which typically resides on endpoint devices) is a component of DNS that application programs access when attempting to resolve domain names to IP addresses (e.g., when accessing a website from a laptop). The stub resolver serves as an intermediary between the application requiring DNS resolution and a recursive DNS resolver. We mention it in this overview because, as we'll learn later, certain attacks target stub resolvers, so it is important to know their role.

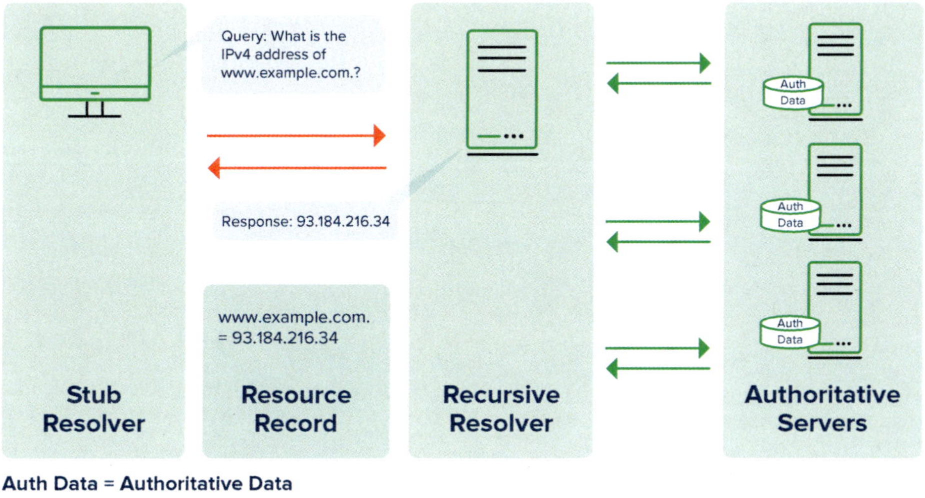

Figure 3: DNS Components

Figure 3 represents the overall resolution process. It begins with the stub resolver, which sends a query for a name to the recursive resolver. If the queried name is not already in the cache of the recursive resolver, the recursive resolver then works its way through the name starting with the right-most label (which happens to be the implied root after the last dot).

From the root server, the recursive resolver is referred to the server for `com.`, which then refers it to an authoritative server for the `example.com.` domain. Finally, the authoritative server replies to the recursive resolver with the answer. As the recursive resolver receives data during each step of the resolution process, it stores that data in its cache for a time determined by the authoritative server—known as the time-to-live (TTL) value. After gathering all the data, the recursive resolver provides the answer back to the stub resolver, completing the resolution.

Summary: Introduction

We want to reiterate the importance of DNS as the first step in nearly all network communications. DNS takes place before communications such as HTTPS, SMTP, and many others. Readers with an inner Threat Hunter may already be thinking of leveraging information in DNS to uncover threats, and they would be right. We will discuss more about how to do so in Chapter 4 and Chapter 6, as well as some of the challenges in analyzing DNS data in Chapter 3 and Chapter 8.

CHAPTER 2

DNS and Malware

As noted in Chapter 1, more than 90 percent of malware uses DNS in some form. DNS provides the avenue for contacting C2 servers that download malicious payloads onto victims' devices. In some cases, malware even makes use of DNS to avoid detection. In this chapter, we will explore the interplay between DNS and malware.

Cyber Kill Chain and DNS

If you are reading this book, you are probably already familiar with the Cyber Kill Chain model developed by Lockheed Martin. Let's quickly refresh the seven steps of the kill chain model, and briefly discuss where DNS fits in the picture:

1. Reconnaissance
2. Weaponization
3. Delivery
4. Exploitation
5. Installation
6. Command and Control (C2)
7. Actions on Objectives

Reconnaissance

In this step, attackers harvest whatever information they can get about the target, such as email addresses, IP information, and products in use. DNS, as a publicly viewable database, is a prime target in this step. By probing DNS, attackers can learn:

- Mail server names and addresses
- Address spaces, sometimes both public and private ones
- Host names, which may give away what system or software is running (e.g., `exchange01.example.com`.)
- Active Directory (AD) domains, which can often be discovered from misconfigured clients, internal domain name—`company.local.`, for example.
    - AD domains also may be visible in the cache of publicly accessible recursive resolvers (aka open resolvers) on the Internet because clients outside of their home network still send queries for resources on their AD domain.
    - In the case of a shared namespace between internal and external networks (common in academic environments), it is easy for an attacker to query for common names in subdomains, such as `_tcp` (used by AD environments), to see if they resolve and are hence potentially exploitable.
- User behaviors and popular domain destinations from recursive queries (e.g., users frequently visiting `example-vendor.com`.)

Weaponization

Typically, attackers leverage software and hardware vulnerabilities to launch malicious payloads. DNS does not play a role in the creation of those payloads, only in their spread and execution. Where weaponization does come into play is through attacks that corrupt DNS servers or leverage open resolvers to generate traffic in a DoS attack. The focus of this book is on abuses of the DNS protocol, not the underlying DNS infrastructure (i.e., the DNS servers themselves). There are plenty of resources available should you wish to learn more about how to secure the many flavors of DNS servers.

Delivery

Attackers deliver the weaponized bundle to the target device, most commonly via the web (HTTP or HTTPS), but increasingly over DNS, and sometimes over encrypted DNS (discussed in detail in Chapter 8). No matter which channel attackers choose to deliver the malicious payload, it always relies on being able to look up the domain name first. *This places DNS at the forefront of malware delivery.*

Exploitation

Attackers exploit a vulnerability to execute code on the victim's system, such as a vulnerability in the email server software. While this step does not directly involve DNS, the DNS service itself can also be exploited. For example, the attacker could take control of the authoritative DNS server by exploiting a vulnerability in the DNS software. We will not discuss these exploits in detail because they usually focus on specific DNS server implementations, which is beyond the scope of this book. Instead, we will focus on how DNS is being abused to exploit other services.

Installation

Attackers install malware on the asset. In many cases, this step may not directly involve DNS. The malicious payload may have been downloaded or transmitted via DNS, but the actual local installation usually happens without involving DNS, by relying on techniques such as privilege escalation.

Command and Control (C2)

Once devices have been compromised, attackers establish a command channel for remote manipulation of the victim's devices. Increasingly, these communication channels run over the DNS protocol. A detailed discussion and case studies on this topic appear in the next section of this chapter.

Actions on Objectives

Attackers once were content to simply degrade or disrupt DNS services, which would in turn take a network down (e.g., via a DDoS attack). This goal still exists, but today's attackers have learned that they can inflict far more damage by using DNS to help spread or disguise malware and ransomware as well as to exfiltrate valuable data or infiltrate malicious code.

As we've just seen, while DNS does not directly come into play in every step of the cyber kill chain, it is integral to most of them. Now, let's take a closer look specifically at the relationship between malware and DNS.

How Malware Uses DNS

Most malware infects target devices with a small executable, then contacts C2 servers to determine what to do next. Because most existing perimeter defenses, such as web proxies and firewalls, closely inspect HTTP/S for the presence of malware, threat actors are increasingly turning to the less scrutinized DNS channel as their pathway of choice. In this section, we look at two case studies that illustrate how threat actors are (mis)using the DNS pathway:

- To determine whether to remain dormant or whether it is safe to launch an attack (e.g., WannaCry)
- To find an HTTP server from which to download more executables (e.g., Black KingDom)

Malware also uses DNS to download malicious executables via DNS tunneling, which is a topic we cover in detail in Chapter 5.

Case Study: WannaCry

```
Classification: Ransomware, Worm
AKA: WannaCrypt, Wana Decrypt0r, Wanna Decryptor
Active: Since 2017
```

The ransomware known as WannaCry has been observed since the year 2017. It exploits weaknesses in Microsoft's Server Message Block (SMB) protocol to infect the victim, encrypts files and then spreads by infecting other devices on the network. WannaCry is devastating ransomware because it not only encrypts valuable files on the victim's device but also spreads like a worm, the first major ransomware to do so.

WannaCry is an example of how malware can use DNS to assess whether conditions are right to proceed. Before launching its attack, WannaCry attempts to resolve a specific domain name,[9] one that attackers know does not exist because they made it up. Based on the outcome of that resolution attempt, the malware makes the following determinations:

- If the domain name resolves successfully, the ransomware assumes that it is in a sandbox environment, so the ransomware lays dormant.
- If the domain name does not resolve, the DNS servers return the status code NXDOMAIN (short for Non-Existent Domain). If WannaCry receives NXDOMAIN for this lookup, it assumes it must be on the real network and is now free to wreak havoc.

This sandbox detection feature was most likely put in place intentionally by the malware author to avoid detection. Many security researchers study malware in a sandbox environment where all DNS queries are trapped and receive forged answers. WannaCry is written specifically to identify and respond to this defensive tactic. The malware looks up a hard-coded gibberish domain that should result in NXDOMAIN. If it receives any response other than NXDOMAIN, the ransomware assumes it is being watched in a sandbox and stays quiet to avoid detection.

Fortunately for system administrators, a security researcher discovered this behavior and realized that the domain name was not only hard-coded but also unregistered. He registered the domain name and while that did not help devices that had already been infected, it slowed the spread and bought time for administrators to address it. This approach became known as the "DNS kill switch"—by changing or stopping DNS lookup results, it changed or stopped the malware's behavior.[10]

Case Study: Black KingDom

`Classification: Ransomware`
`Active: Since March 2021`

First discovered in early 2021, the Black KingDom ransomware infects devices through Microsoft Exchange Server ProxyLogon vulnerabilities. Once installed on the target, it runs a PowerShell script to download additional ransomware executables from the hard-coded domain `yuuuuu44.com.` via HTTP.

Traditional blocking mechanisms, such as firewalls, have historically operated by blocking IP addresses, which is ineffective against Black KingDom because the domain name would resolve to different IP addresses over time. This technique

9 The domain name www.ifferfsodp9ifjaposdfjhgosurijfaewrwergwea.com, which is no longer in use today.

10 An account of the discovery by Marcus Hutchins and Jamie Hankins, the registration of the kill switch domain by Hutchins, and what happened afterward can be found here: https://techcrunch.com/2019/07/08/the-wannacry-sinkhole/

of cycling through multiple IP addresses, known as fast flux, makes it difficult for traditional security blocking mechanisms to keep up with ever-changing malicious IP addresses.

The good news: Because the domain that Black KingDom uses is known and does not change, it is easy to block with DNS. By using DNS, you can block it once and you're done. This blocking can happen in one of two ways:

- *Resolution blocking:* If the target device cannot resolve the name `yuuuuu44.com.` to an IP address, it cannot make the subsequent HTTP connection to download additional ransomware components, which stops the threat from causing any more damage.
- *Response substitution:* Administrators can replace the malicious IP address in the DNS response with an IP address of their choosing, such as the IP of a server run by the security team. This option has the added security benefit of enabling administrators to use connections to the server to identify all devices infected with Black KingDom.

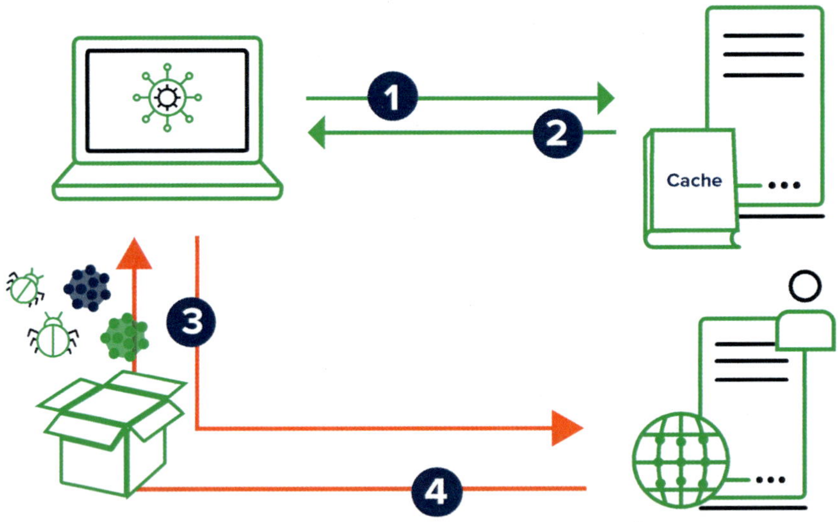

Figure 4: Black KingDom illustrated

1. Infected victim looks up hard-coded domain `yuuuuu44.com.`
2. DNS server responds with the IP address(es) of `yuuuuu44.com.`
3. Victim makes HTTP connection to attacker-controlled server
4. Victim receives additional malware component(s)

Summary: DNS and Malware

As we have seen, some malware uses DNS to evade detection, and nearly all malware uses DNS to look up the destination for its C2 communications. If administrators are blocking based on IP addresses only, attackers can easily bypass the blocking mechanism simply by pointing to a different address in the DNS resolution.

For example, attackers can have the domain name `dwnld.attckr.com.` resolve to 17.17.17.17 today and 18.18.18.18 tomorrow, making it difficult for security administrators to keep up.

What we have learned so far:

1. Malware uses DNS as a probing technique to avoid detection
2. Malware uses DNS to receive instructions on next steps
3. Blocking DNS lookups or rewriting responses can stop malware from executing attacks

CHAPTER 3

Look-Alike Domains

What Are Look-Alike Domains?

Look-alike domains are domain names that appear to be nearly identical to well-known domain names; they are also known as homograph attacks. It is also common to see the term written without the hyphen (i.e., lookalike domains). Look-alike domains are almost always malicious, and their use can range from typo-squatting, which exploits common typos in domain names, to deliberate domain name impersonation.

Today's cyber criminals register hundreds of thousands, and sometimes millions, of domains and use them to carry out phishing attacks or as part of a business email compromise (BEC). In this chapter, we look at how threat actors abuse look-alike domains to gain the trust of end users and to hide their tracks from cyber-security investigators. We also look at several similar threat tactics that work by obscuring domain names. We will get into the technical details later, but let's start by looking at an example:

Case Study: PayPai.com.

```
Classification: Look-alike domains
Observed: 2000
Tactic: Impersonation of popular payment sites; one of the
first documented cases of look-alike domain and phishing
technique
```

The year was 2000. The tech world was still basking in the glory of surviving Y2K, congratulating itself for averting possibly the most over-hyped technological disaster known to mankind. But 2000 also marked another historic event in cyber security: the first well-known DNS impersonation attack.[11]

A scammer registered the name `paypai.com.` (replacing the "l" character at the end with an "i") and set up a web page that looked exactly like the legitimate site. He sent an email to PayPal users notifying them they had money coming their way from PayPal and to click on the link to receive their big payment. The link pointed to `www.paypai.com.` as the destination. To disguise the lowercase letter "i" at the end of the URL, the scammer used an uppercase "I" but in a smaller font so it would look like a lowercase "L" ("l"). When victims went to the web page, they were prompted to submit their credentials, which the scammer stole.

This was the first of many similar impersonation techniques that would come to be known as "phishing."

---

11 The original ZDNet article from the year 2000: https://www.zdnet.com/article/paypal-alert-beware-the-pay-pai-scam-5000109103/

Look-Alike Domains Today

Today, scammers use a variety of look-alike domains to trick people. Below are some examples of look-alike domains for `gmail.com.`, the popular Google email service. See if you can spot how each one differs:

- `gmial.com.` (transpose letters "a" and "i")
- `g.mail.com.` (subdomain subversion)
- `gmaill.com.` (repeat letter "l" twice)
- `gmai1.com.` (substitute letter "l" with number "1")
- `g-mail.com.` (hyphen between letter "g" and "m")
- `gmail.co.` (different top-level domain, "`co`" instead of "`com`")
- `gnnail.com.` (two letter "n" in place of letter "m")

Many users will not pay close attention to these subtle differences, especially if the web page or email looks legitimate to the eye. Many web pages used in look-alike domain scams also have valid TLS certificates. Attackers can obtain those certificates because they own the false domain. The use of these certificates increases the chance that victims will be fooled.

Business Email Compromise

Deceptions that rely on look-alike domains are not limited to web pages. Misleading domains are also used in exploits that impersonate legitimate business email addresses. According to the FBI's Internet Crime Complaint Center, BECs have been growing year after year.[12] Financial damage or loss to businesses has grown from $360 million in 2016 to over $1.8 billion in 2020.[13]

The most common BEC tactic is for the threat actor to register a look-alike domain and send the would-be victim email communications. In 2019, a Chinese venture capital (VC) firm sent $1 million in seed funding to someone it believed represented an Israeli startup.[14] There was a real Israeli startup that needed the seed money. However, the fraudster registered look-alike domains, by adding the letter "s" to the end of the legitimate ones, and inserted himself, via email communications, between the Chinese VC firm and the legitimate Israeli startup, in classic man-in-the-middle (MITM) style.

---

12 https://www.ic3.gov/Home/AnnualReports

13 https://www.ic3.gov/Media/PDF/AnnualReport/2020_IC3Report.pdf

14 https://research.checkpoint.com/2019/incident-response-casefile-a-successful-bec-leveraging-lookalike-domains/

Every email sent by the VC and the real startup went to the attacker first, who then reviewed and edited the message, before forwarding it to the intended recipient. In the end, the fraudster sent 18 emails to the Chinese VC firm, 14 to the Israeli startup, and walked away with $1 million.

International Characters in DNS

At this point, you might think: I am not a n00b! I read email addresses carefully, I am fluent in 1337, and I can certainly tell the difference between google and g00gle in my web browser.

Think again.

The original design of DNS accounted for characters only in ASCII, thus excluding languages that use non-ASCII characters, such as Arabic, Greek, and Japanese. That changed in 1998 with the adoption of a new encoding standard, Punycode,[15] which allowed Unicode characters to be represented using the 63 ASCII characters[16] allowed for hostnames. This made it possible for people to register domain names in any language.

Figure 5 shows an example of Unicode (left) and Punycode (right). The top domain name is "panda" written in Chinese; the bottom domain is "vodka" written in Russian.

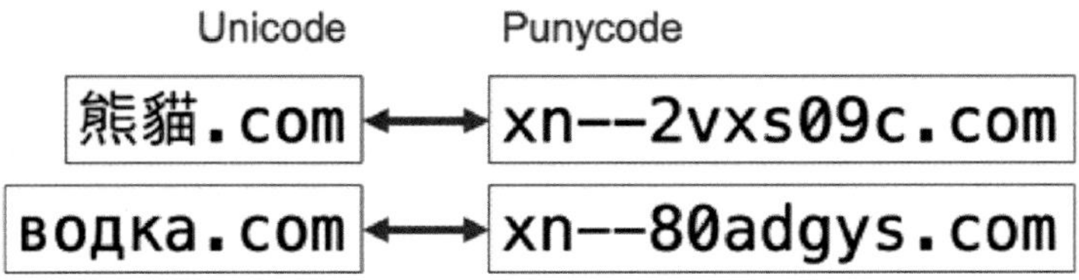

Figure 5: Punycode examples

While internationalized domain names (IDNs) were a necessary step for the globalization of the Internet, it gave scammers and fraudsters everywhere more options to trick their victims. There are many characters in different languages that look very similar. In some cases, they are outright identical.

15 The Punycode specification is documented in RFC 3492, ratified in the year 2003. Today, there are many tools available to convert between Unicode and Punycode.

16 Supported characters are a-z, A-Z, 0-9, and hyphen. See Chapter 5 for more details.

Using IDN for Look-Alike Names

Below are three versions of the `PayPal.com.` domain name. Can you spot the differences?

- `PayPal.com.`
- <u>`PąyPąl.com.`</u>
- `PаyPаl.com.`

The first domain name is the normal `PayPal.com.` domain name. Other than the capitalization (which is allowed because DNS is not case sensitive), there is nothing wrong with this domain name. It only uses standard English (Latin) letters.

The second domain name replaces the letter "a" with the Polish letter "ą," which adds the ogonek tail-shaped accent mark to the letter's lower right (Unicode 0x105). To further hide the difference, the attacker uses the underline, disguising it as a clickable hyperlink, at the same time hiding the visual differences. The Punycode encoding is `xn--pypl-btac.com.` for this domain.

The third domain name replaces the letter "a" with the Cyrillic letter "а" (Unicode 0x430). This one is even more difficult to detect because the two characters look so similar that most people cannot tell them apart. The Punycode encoding is `xn--pypl-53dc.com.` for this domain.

While there is some hope that keen observers could spot the very subtle differences in the examples above, the following example domains use characters that are so similar, nearly no humans can detect their differences. Study the following three domain names carefully and see if you could identify which one is the real `google.net.` and which are the impostors.

- `gооgle.net.`
- `gοοgle.net.`
- `google.net.`

The first domain name replaces the letters "o" with the Cyrillic letters "о" (Unicode 0x43E). The second domain name replaces the letters "o" with the Greek letter "ο" (omicron) (Unicode 0x3BF). Only the third domain name uses all English (Latin) characters.

By now it should be very clear to you, there are simply too many of these subtle visual variations for any security professional to identify and block one by one. We will show how to leverage existing threat intelligence to block many of these domains later in this chapter.

New Top-Level Domains

Traditionally, there were only a handful of top-level domains (TLDs); these included well-known domains, such as `com.`, `net.`, and `org.`, and the country-code top-level domains (ccTLDs), including `ca.`, `in.`, `jp.`, `mx.`, and `us.`, among others.

As if the DNS namespace was not convoluted enough, in the year 2012, the Internet Corporation for Assigned Names and Numbers (ICANN) opened the floodgates by accepting applications for new TLDs. As of the year 2021, there are about 1,500 TLDs,[17] compared to just over 300 in the year 2010. Names such as `dev.`, `gmail.`, `feedback.`, `lease.`, `pizza.`, and 我爱你. (Chinese for "I love you", Punycode `xn--6qq986b3xl.`) are all real and resolvable TLDs.

What is worse, not all new TLD entities uphold the same standards and policies when it comes to name registration. It probably doesn't come as a surprise that the company Donuts, Inc.,[18] which registered the TLD `pizza.`, may not have the same checks and policy enforcements as the company that maintains the TLD `com.`, Verisign. In 2020, a curious user tested the limits and was able to register the following domain names[19] for himself:

- `facebook.网站.`
- `microsoft.みんな.`
- `netflix.soy.`

As the number of new TLDs grows, it is difficult to keep up with which ones are legitimate and which are not. Take the following two real domain names, for example:

- `google.한국.`
- `google.קום.`

The first domain name's Punycode is `google.xn–3e0b707e.`; the TLD is "Korea" written in Korean. This domain is registered by MarkMonitor to prevent abuse. The second domain name's Punycode is `google.xn–9dbq2a.`; the TLD is "`com`" written in Hebrew. This domain is registered by an individual, the same curious user who also registered the `netflix.soy.` domain.

17 For a complete list of TLDs, visit https://data.iana.org/TLD/tlds-alpha-by-domain.txt

18 This is not to pick on Donuts, Inc.; it is just an example to show how easily this information can be looked up using tools such as WHOIS.

19 https://tinyprojects.dev/posts/i_bought_netflix_dot_soy

Related Techniques

Even without IDNs or new TLDs, threat actors can still trick users into believing the legitimacy of their domain. We will briefly discuss four techniques, the first and third of which target mobile users, whom attackers safely assume have smaller screens, the second takes advantage of a little-known fact about domain labels, and the final technique targets browsers overall.

Partial Matching Names

Mobile devices, such as smartphones and tablets, have smaller displays than desktops and laptops. Less space is an opportunity for fraudsters who use long domain names, such as:

- `app.downloads.microsoft.com.attacker.org.`
- `content.streaming.netflix.com.attacker.org.`
- `users.shopping.amazon.com.attacker.org.`

Chances are, users on smaller screens will never see the full domain name. As was mentioned earlier, the attacker can obtain TLS certificates for the domain (`attacker.org.` in our examples) he or she controls and proudly send the certificate to the unsuspecting user. Users will see that coveted lock symbol in their browser or application and proceed with full confidence that they are safe from harm, only to step into the attacker's trap.

Hidden Path

Another potential technique that attackers could use is one we'll call the "hidden path" technique.[20] It consists of hiding the target domain by structuring subdomains in such a way that most people would consider the target domain to be merely path information. Most people believe the 63-character set for DNS hostnames that was mentioned in the IDN section above applies to the entire FQDN, but it does not. Within the domain labels, any character is allowed, including the / character (U+002F), and some Unicode characters that look like forward slashes (U+2044 and U+2215). An attacker could take advantage of the support for those characters coupled with the general lack of awareness that they can be used in a domain name. One particular way is what we describe here, another is discussed as a part of Chapter 5, "DNS Tunneling."

The / character is commonly used in URLs to designate a path. It is so common that most people presented with a FQDN that contains a / likely wouldn't realize it is a part of the domain name at all. So, an attacker can use a subdomain containing a / to hide the actual domain that is in use. The easiest way to

---

20 At this time, we are not aware of any active use of this technique, but we want to make the security community aware of it in case threat actors should begin to use it.

understand how this can be done is through an example. Let's start with the following:

`google.com/path.attckr.com.`

Most people looking at that string of text would believe it is pointing to a resource on a server owned by Google, believing that "`path.attckr.com.`" is merely path information. However, in reality, the domain is "`attckr.com.`" and there is a subdomain in it called "`com/path`." Within that subdomain is a resource record called "`google`" that the attacker would configure to point to its own server. Multiple layers of subdomains containing / could be used here to further disguise what is going on, making it look even more like path information.

The use case for this with the normal / character (U+002F) in a web browser is most likely limited, because the browser would probably treat everything after the / as a path and would resolve the name before the slash, in this case, `google.com.`, as the target host. But it would be easy for malware to use such a domain name to hide in plain sight when conducting C2 communications or exfiltrating data.

The existence of unicode characters such as U+2044 and U+2215 that look nearly indistinguishable from a normal forward slash are where the real danger lies. A subdomain containing a U+2044 or U+2215 would appear as path information to the victim, even though a web browser would not process it that way. This would make it simple to create clickable links in a phishing email that appear to be from a legitimate domain, especially when lesser-known TLDs are used. Before being made aware of this, if you saw such a name in an email or DNS query log, would you have considered it a potential threat, or would you have believed it to be traffic destined for Google and moved on?

URL Bar Masking

Some attackers just eliminate domain names from the picture, by using various tricks that hide the URL display bar from users. If users can't see the domain name, the only way they can judge the trustworthiness of the web page or application is by its appearance.

One such attack method demonstrated in 2019 is known as Inception Bar.[21] In this method, the attacker takes a screenshot on the mobile device (the proof of concept used an Android phone) and uses the URL bar image as part of the web page. For example, when successfully executed, users visiting `www.attckr.com.` will see the impersonated content for `bank-site.example.com.`, along with the screenshot of the legitimate URL displayed near the top.

---

21 https://jameshfisher.com/2019/04/27/the-inception-bar-a-new-phishing-method/

Malicious Overlays

HTML has come a long way since the late 1990s. Its capabilities have grown to the point where entire remote desktop access is possible with just HTML5, allowing users to remotely control desktops graphically with nothing but their web browsers. But these new powers open new possibilities for bad actors as well. For example, they can use native HTML/CSS to create visuals that appear to be legitimate web browser windows, pop-ups, and buttons, tricking unsuspecting users into divulging information.

One such example is the browser-in-the-browser attack,[22] where a hacker prompts unsuspecting users to provide login information to other legitimate websites with illicit prompts that appear visually identical to the legitimate ones (see Figure 6[23]). For example, a user is first tricked into visiting a nefarious website through a bad link. The site looks genuine. Once they arrive, a pop-up window prompts the user to login to the `facebook.com.` site. The pop-up window is an exact copy of a real Facebook login screen that attackers created using code such as HTML5 or JavaScript. This type of attack has been around for decades, but threat actors are finding new ways to use it to trick users into sending information to malicious domains while hiding the real destination.

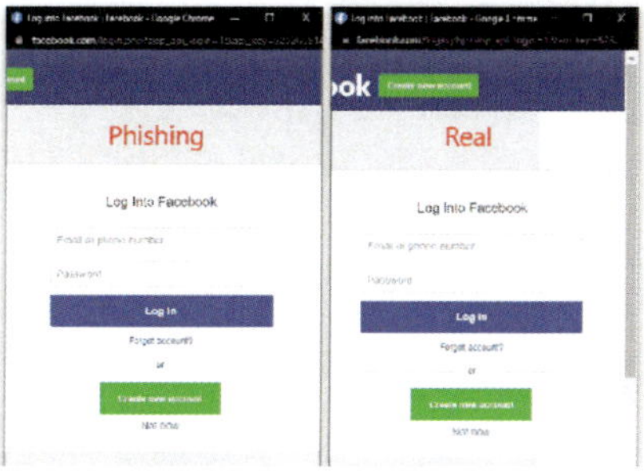

Figure 6: Browser-in-the-browser deception

Common Characteristics

We have seen five attack techniques that rely on manipulating domain names to obscure bad web destinations: look-alike, partial matching, hidden path, URL bar masking, and malicious overlays. Let's take a brief intermission and focus on what all these techniques have in common: static domain names. With look-alike domains, the threat actor needs to register a few different domain names that appear to be visually similar to the impersonated one. With the related four techniques, the domain name is obscured, so attackers do not need to acquire multiple ones. They can simply reuse or recycle existing illicit domain names and save a few dollars in domain name registration fees at the same time.

22 https://github.com/mrd0x/BITB

23 Image credit: https://mrd0x.com/browser-in-the-browser-phishing-attack/

At this point, you may be wondering: If these attacks rely on static domain names, why can't we just flag the associated IP addresses they resolve to and block those addresses with firewalls and other perimeter defenses? The problem with IP address blocking is that a single domain name can resolve to a multitude of different IP addresses. As a result, IP address blocking is far less effective than in years past. Attackers have learned to evade that defensive tactic by changing IP addresses frequently. We will see more examples of the ineffectiveness of IP address blocking in Chapter 4.

From a security standpoint, blocking based on domain names can be much more productive. The good news for security practitioners (for now) is that with these particular techniques, individual attackers are using relatively few domain names (because they want to keep domain registration costs low). That means it is possible for defenders to block these domain names manually. The bad news? There are *a lot* of malicious actors in the world, so the number of malicious domain names is soaring. As a result, manually adding rules to block the endless torrent of bad domain names is not sustainable.

What Can We Do?

At the root of the problem is the ability for anyone to register nearly any domain name they want, with very few restrictions. This freedom at one point was the appeal of the Internet, but from a security perspective, it has become a monster that we can no longer control. Organizations such as the DNS Abuse Institute[24] are trying to rein the monster back in, but this effort will take time. Meanwhile, automated blocklists are our best option.

Response Policy Zone (RPZ)

As we've just seen, cyber criminals have many ways to create domain names — look-alike names, new top-level domains, and names with international characters. But wait, there's more, attackers can also use a time-based algorithm to select different names for the same malicious asset on any day (see Chapter 4). Consequently, knowing which names to block on a particular day becomes a monumental task.

You might be thinking: Great, I am already short on time, and now I discover a whole new area that is literally a full-time job to maintain! No worries, the Response Policy Zone (RPZ) has your back. RPZ is an open standard[25] that defines a special DNS zone type that contains security policy actions mapped to domain names. An RPZ captures good domains as well as malicious ones. When in use, an RPZ allows DNS queries to flow unimpeded to legitimate domains while blocking access to malicious ones.

---

24 https://dnsabuseinstitute.org/

25 RPZ was designed by Paul Vixie and Vernon Schryver in 2009. Many thanks are owed to them for this incredibly useful security tool. See https://datatracker.ietf.org/doc/draft-ietf-dnsop-dns-rpz/

Information about newly identified unlawful domains (as well as new legitimate ones) can be continually added to an RPZ. This updated information can be readily shared for enforcement purposes because RPZ is a DNS zone file, which means that the policies the RPZ contains can be acquired by recursive name servers using DNS zone transfers. As new domains are added to the policy, updates can be published using the built-in zone transfer mechanism over port 53.[26]

Using the policy actions defined in the RPZ for given domain names, recursive servers can then decide, on a query-by-query basis, how to respond. You can build your own local RPZ policies or receive them via zone transfers (known as RPZ feeds) from your favorite security vendors.[27] Your recursive name servers will know about malicious domains and how to handle clients when they send queries for malicious domain names. Figure 7 illustrates how RPZ works at a high level.

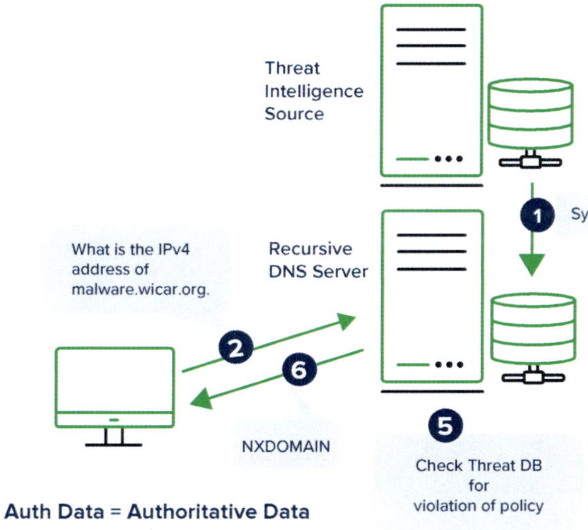

Figure 7: RPZ Overview

Step 1: Recursive name server pulls the threat database from security vendor via TCP port 53

Step 2: Client queries for a domain name

Step 3 - 5: Depending on the exact configuration and policy, the recursive name server may query the target's authoritative name server, which malicious actors may own, to retrieve the answer and then check for policy violation

26 Zone transfers commonly use TCP port 53, but it is possible with a small-enough amount of data to perform incremental zone transfers via UDP port 53.

27 There are also community-provided feeds, available via https://pi-hole.net/

Step 6: Recursive server responds to the client with the answer configured by the policy; the most common response is NXDOMAIN

Common RPZ Policy Actions

The RPZ specification includes many possible actions, from PASSTHRU (allow query to go through) to NODATA (returns no data in the response). The three most common actions include:

- Block by name [NXDOMAIN or NODATA]: "You are looking up a malicious name. Don't!"
- Block based on response IP address [NXDOMAIN or NODATA]: "The answer to your query contains malicious IP addresses. Don't go."
- Redirect [CNAME]: "You are looking up a malicious domain. I will redirect you to something else."

RPZ Action: Block by Name

This is perhaps the most common action configured in RPZ. When a user queries for a "bad" domain name, the recursive server responds with NXDOMAIN (or other similar responses). Remember, the NXDOMAIN response is an authoritative server telling the querier that the record requested does not exist. Because the client is unable to resolve the name and contact the malicious destination, it is thus protected from the malicious content. Depending on the application, users will likely see generic error messages such as "This site can't be reached" or "Name not found," or sometimes there may be no error message at all. Figure 8 shows a simplified version of how the blocking works:

Figure 8: RPZ block by name

1. Client queries for the name `www.newdomain.com.`
2. Recursive name server checks against local RPZ data for policy violation.[28] There is a policy that blocks anything in `newdomain.com.`, and the action is to respond with NXDOMAIN
3. Recursive server sends back NXDOMAIN to client

---

28 The order of RPZ enforcement versus recursion is controlled by how the DNS server is configured; it can check either before recursing or after based on the setting for qname-wait-recurse. Many implementations default to checking for violation after retrieving answers, which is different from our simplified illustration.

RPZ Action: Block Response IP Address

Sometimes we know which IP address the malicious actor is using, but we are not sure what new domain name(s) maps to that address. It's worth remembering that IP address-to-domain name mapping is not strictly one to one: one IP address could have multiple domain names associated with it and vice versa. RPZ provides a mechanism for blocking the response by looking for the offending IP addresses in the DNS response. This ability to block based on response data works hand-in-hand with query data. In this scenario, the recursive name server must contact the target authoritative server (often run by the malicious actor) to receive a response. That response from the authoritative server contains the malicious IP address, which triggers the block. Figure 9 shows how this works:

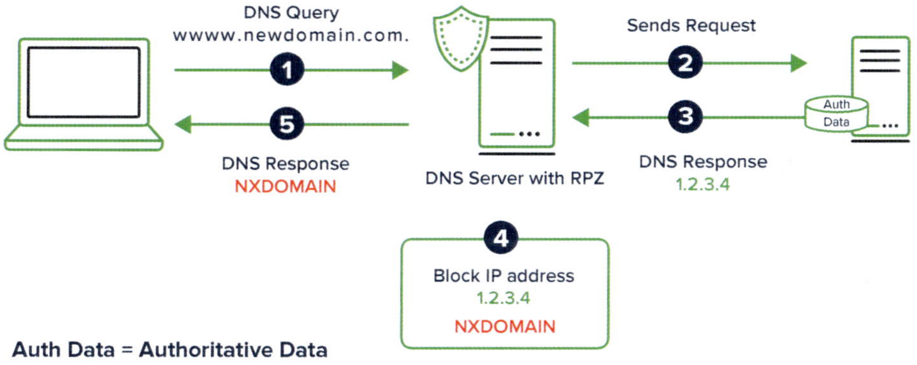

Figure 9: RPZ block response IP address

1. Client queries for the name `www.newdomain.com.`
2. Recursive name server follows normal DNS resolution process to find the authoritative server for the `newdomain.com.` domain
3. Recursive name server receives the authoritative response for `www.newdomain.com.`, which contains the IP address 1.2.3.4
4. The policy blocks the IP address 1.2.3.4
5. Recursive name server sends NXDOMAIN to the client

RPZ Action: Redirect/Rewrite

Sometimes we don't want the clients to just receive an NXDOMAIN. We want the clients to resolve to a different name (and destination), usually some type of captive portal or remediation system. Such redirection is usually deployed along with a web server that we control, one that can serve custom content, such as "You tried to access a malicious domain and we stopped you; click here to learn more."

The same technique is frequently used by Internet hot spots or captive portals at hotels and airports, where customers pay to access the Internet. When we use the CNAME (canonical name) record response in DNS, we basically state, "This domain name points to another domain name." We are then able to redirect the client's traffic as Figure 10 illustrates:

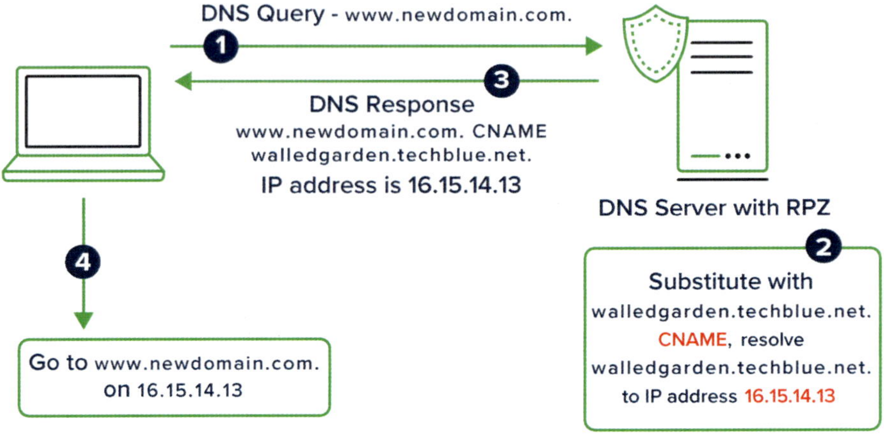

Figure 10: RPZ redirect

1. Client queries for the name `www.newdomain.com.`
2. Recursive name server checks policy, finds this name is in violation, and substitutes or rewrites the name with `walledgarden.techblue.net.`, and resolves the substituted name to an IP address, which is 16.15.14.13.
3. DNS server responds to client that `www.newdomain.com.` is a CNAME to `walledgarden.techblue.net.`, and its IP address is 16.15.14.13
4. Client connects to `walledgarden.techblue.net.`, a domain most likely controlled by the administrator, and some type of information message appears, such as "You cannot access `www.newdomain.com.` because it is against corporate security policy."

Summary: Look-Alike Domains and RPZ

Scammers and malicious actors fool users into accessing fraudulent destinations using look-alike characters or patterns in the domain name. Sometimes they employ tricks that don't rely on DNS itself to make the domain or web page appear to be legitimate. One of our defensive options is to deploy RPZ, either locally configured or as a curated list from a trusted threat intelligence source to automate the task of blocking malicious domains. Our recursive resolvers can then start filtering DNS responses, preventing clients from accessing known malicious domain names.

CHAPTER 4

Domain Generation Algorithms (DGAs)

A domain generation algorithm, more commonly known as DGA, is so named because it relies on an algorithm to automatically generate (and register) domain names. In this chapter, we explore how malware uses DGAs and provide some defense techniques against DGAs.

DGA Basics

DGAs usually fall into two categories: pseudo-random names or, more recently, dictionary-based names. When a dictionary-based name is used, it is called a dictionary DGA or DDGA.

Pseudo-random names look like someone just dragged a cat over the keyboard, resulting in unpronounceable names that appear to be gibberish: `cuxrbykwokrqcldyjawc.com.` and `enjvatxpymxbcddidbmc.com.` are two examples.

Dictionary-based names are often generated by combining two or three words from a dictionary or group of dictionaries to generate domain names that look plausible and potentially created by humans, rather than random strings of gibberish that can trigger detection systems. For example, the first dictionary may contain adjectives like "happy," "sad," and "lovely"; the second dictionary may contain nouns such as "shoes," "cars," and "ring." When combined, they yield domain names such as `happycars.com.` or `sadshoes.net.`, which appear like more normal domain names.

Why Attackers Use DGAs

Individual domain names are relatively cheap to register, but registering hundreds or thousands of them can be very expensive. So why are attackers spending so much money to register a bunch of domain names? Because as a threat tactic, it is very effective—enough to pay for itself.

The reason DGAs are effective is easy to understand—flooding the Internet with vast numbers of domain names makes it very difficult to detect and block them all. The key to how DGAs work is that threat actors do not need every domain name to resolve to the malicious destination. Rather, they are carefully choosing a few select domains that would respond to queries, while the rest of the dummy domains act as smoke screens to hide their tracks.

We will look at two different case studies to see how attackers are using this DGA technique to evade detection and blocking.

Case Study: Zloader

```
Classification: Trojan
AKA: Zeus Sphinx, Terdot, DELoader
Active: First observed in 2015, resurfaced during 2020
during COVID pandemic
Specialty: Sends false COVID-19 information with .doc or
.docx attachment, which executes macro to install the
downloader on the target system. Once installed, downloader
communicates with C2 servers to fetch relevant exploits and
malware.
```

Zloader is a family of trojans that has evolved over years. Here we want to focus on a particular version seen during 2020, during the peak of the COVID-19 pandemic. This version uses over 30,000 domain names[29] and a time-based algorithm to rotate through the domain names. For example, Figure 11 shows the 32 domain names used for the date December 5, 2020:

```
exxkosgrtkidhlibffoi.com.     klclxchkitlnsyxmsbgl.com.
kbucdfkqpeksffoikkgb.com.     smuaqqnjfakdwjoadorp.com.
kiadqpfykieuudikvhxd.com.     ttlkaxpscaryjfwagwpe.com.
nctdftcmhovdaoktsdag.com.     gbycqfkprotmaahctctc.com.
pwipbixsquumlfexktsy.com.     poqqepsalwaxcelcpara.com.
mdexkspytmskcamincqd.com.     gwpegbajewhmxbyhnaoy.com.
ilavdthjpgjkjsbroueh.com.     ukjsbrjqtrlsffmcoxqv.com.
broqmmuvglsidfnsapky.com.     urnecwobsoapwrjkpoto.com.
vxthjgydpejomvcsgsxy.com.     xqvubbiqdfqtfbqdijxi.com.
jgydyuxdtxpgvpkpckhq.com.     acoyqsfphjvgxaftyoci.com.
cwhmkcwdvghfvkyafyxn.com.     pnbqbgvyipopygcvevik.com.
hqcnmgfdyhtmvmeyapwp.com.     bhxnhevinycwgsleikru.com.
qgngtxpgcleheqioocfw.com.     sjjwiiljdtiobdtfhavm.com.
vhwhpvysydaswwgjteew.com.     bxpsiqqrwsiatkymskwk.com.
ttxdeqrthgiimagdehbr.com.     jseyjsiqeynghffbnkyj.com.
xlsvcdmxnrauadbulpev.com.     pgiltojbclwadrpwvgpa.com.
```

Figure 11: Zloader domains for December 5, 2020

The attacker may choose to use any of these domain names on that day. Although all these domain names might resolve to some IP address, it could be that only one IP address is actually working on that day. Alternatively, the attacker could rotate through the list during the day.

---

29 https://github.com/baderj/domain_generation_algorithms/tree/master/zloader

The next day, December 6, the attacker would abandon these 32 names and move on to the next set (see Figure 12):

```
rvpidccqxpmugpdnrqjf.com.
mnnprxrobvymbodunumr.com.
bppstmqctiltbwltxski.com.
peapnqrgsnqeskmkmxem.com.
huvvpidccqxpmugpdnrq.com.
jfmnnprxrobvymbodunu.com.
ekrnxjwxcigqpuvvxaga.com.
wjuvvofarpwldfanosqo.com.
yompmepvyvbxkjvydsxt.com.
qdcsjmegbshkoixvpfyq.com.
nvhnqqwmfjriipevojby.com.
nmmbinvhnaicmttrgfrm.com.
llwyxdeaonimwifeesjm.com.
ixvyafwphbujqifnpgfj.com.
lftfkomseacmttrdbmbo.com.
qtyrwphbuvvxadrvpqys.com.
dwlwgqmkmkacvkofnyis.com.
fferknjcovaxihpfbdnu.com.
irjkguthjmbfvntfkkux.com.
xegetjlfktpeyxgvjcfb.com.
tgwrmyklxfscdpwktyyj.com.
vvuikwnfuerbsdthjmrb.com.
stpepyqbnndsygxupgfj.com.
lfkxdhftorftofaqpxnb.com.
twwdokmwebxkvbblobvy.com.
ywdoxlyijrlvtvtmygkv.com.
rtbtgnbkguttimdyndvm.com.
ugpdaqtlqqklhshhkydg.com.
cuhogdxonrtmppvlbdkm.com.
tdynmqsdgcuxxemhusrb.com.
pswqjiecbkyytyvbxkvb.com.
xhrqdptjokqoyoptjogm.com.
```

Figure 12: Zloader Domains for December 6, 2020

The observed behavior for Zloader is that the infected device will contact several domains throughout the day. Many of these contacts will result in an NXDOMAIN response, and some will resolve successfully to IP addresses. The infected device will then use those IP addresses for that day for C2 communication.

As you can see, this type of attack makes a security practitioner's job much harder. Even if you were able to figure out which domain or domains are used for one day and block the associated names and IP addresses, the next day you must start all over again! With DGAs, attackers have the upper hand: they always know exactly which of the thousands of domain names are active at any given time, while defenders are completely in the dark.

Blocking DGAs

It might be tempting to think that RPZ will solve the problem of rotating malicious domain names. Unfortunately, it does not. Don't forget that a technology like RPZ essentially functions as an automatically updatable blocklist, which means each recursive DNS server needs to store the blocklist in memory and compare each outbound query to the list.

This situation is not a problem when there are only, say, 32 names on the list (as with Zloader): Checking a queried domain name against 32 names before forwarding it is unlikely to cause any significant slowdown. However, as our next case study will show, the list of malicious domains has grown out of control.

Case Study: Conficker

```
Classification: Worm
AKA: Downup, Downadup, Kido
Active: 2008 to present
```

The original Conficker worm first emerged in 2008 and infected over 15 million computers. By 2020, there were an estimated 500,000 devices infected with Conficker variants. One of the most challenging aspects is infected legacy and Internet of Things (IoT) devices that depend on old versions of Windows and that make it difficult or impossible to install monitoring agents or anti-virus software.

What sets Conficker apart from other malware is in the massive number of domains that it and its variants generate using DGAs. The earlier versions of Conficker generated 250 domains, while later variants could generate up to 50,000 domains. Infected devices look up the DNS domain(s) and contact the C2 server over HTTP/S.

Here is a small sample of domains that Conficker used on February 12, 2009:

- `tvxwoajfwad.info.`
- `blojvbcbrwx.biz.`
- `wimmugmq.biz.`
- `fwnvlja.org.`
- `umgrzaybbf.ws.`
- `btgoyr.cc.`
- `zboycplmkhc.cc.`
- `qsqzphbn.biz.`
- `xqdvmavs.cn.`
- `wgrrrr.biz.`

Recent studies[30] show that only a fraction of the huge number of domains registered by Conficker are still in use. However, we are uncertain exactly how many are dormant and could be active the next day at the command of a threat actor. As we see in the case of Conficker, one malware alone can use tens of thousands of domain names. Identifying all these domain names is a big challenge, and so is loading this giant blocklist on a DNS server that may already be under heavy load.

---

30 https://circleid.com/posts/20210831-what-are-the-internet-domains-connected-to-the-conficker-botnet

If we load all the Conficker domain names into our recursive resolver, that could be millions of domain names. That means, we need to load this blocklist of potentially millions of domain names into the DNS server's memory, and every outbound query must be checked against each of the millions of names. This method will likely cause a huge DNS performance issue. And since all network connections start with a DNS query, this translates to a general network performance issue.

We need a new approach. And this new approach requires treating domain names as "guilty until proven innocent."

Newly Registered/Observed Domains

Think about how firewall rules were managed in the early days of the Internet. The earlier firewall rules were wide open, only blocking a few known malicious ports or IP addresses. As seasoned network security professionals know, this approach did not last long. It is now a standard practice for firewall administrators to block or drop all traffic and only allow specific flows through.

But how do we do this for DNS names? We start by not trusting newer domain names.

Millions of domains (if not more) are registered each day, and at this point we can say with confidence that most of them are malicious. Are there some legitimate domain names? Sure there are. But we don't know how to readily tell the licit from the illicit. Security experts have devised a new and simple (but not quick) way to discover the difference between benign and malicious names: We assume all new domain names are malicious and wait until we have gathered more information to know which ones are benign (safe) and which ones are malicious (not safe).

But wait, for how long? To figure that out, we need to answer this basic question: How long does it take for the security industry to determine if a new domain is malicious? Figure 13 shows the median time for various threat intelligence firms to detect a malicious domain, which is 43.7 hours, or just under two days.[31]

---

31 Source: Infoblox threat research

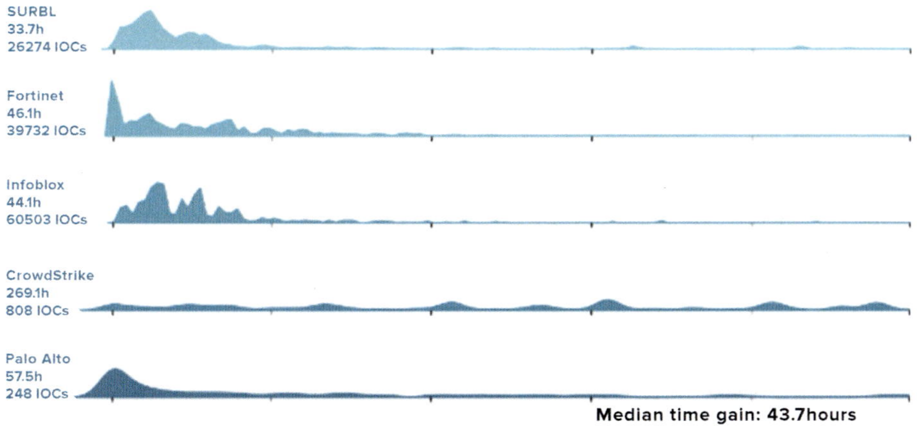

Figure 13: Median detection time for malicious domain name

This means, if I am a threat actor who registered a new domain name, and I am out there spreading my evil DNS seeds, it will take on average 43.7 hours for the good guys to catch on that my domain is malicious and then publish that in their RPZ feed. By the time this new information is propagated and enforced on the DNS recursive resolvers of the world, I have had a run of nearly two days with no one to stop me!

A new category of RPZ feed is derived from this data, and those feeds have different names and settings depending on the security vendor. Some call the feed newly observed domains (NODs), while others refer to it as newly registered domain (NRDs), and still others as newly seen domains (NSDs). These names and definitions differ in very subtle ways, but the basic principle is the same: This list contains new domain names that we are blocking until we know more about them.

Note that vendors define "new" slightly differently. Some go by the new domains registered with top registrars. Some go by new domains that have become resolvable (although they may have been registered a while ago). Some vendors, such as Farsight Security, check against a historical Internet traffic database to determine whether the new domain name is observed as active for the first time.

This new RPZ category allows security administrators to block known malicious domain names (proven guilty) and emerging domains (assumed guilty). In time, some of the assumed-guilty domains will prove themselves malicious and be moved into the proven-guilty RPZ blocklist, while others will wait for the grace period to pass and become resolvable on the Internet.

This approach also cuts down the blocklist size significantly because instead of blocking all possible malicious domain names, we block only the known ones (proven guilty) and the new ones (assumed guilty). It is hard to know how many new domains appear daily. Rough estimates in 2020 were around 600,000 to 1.5 million.[32] This is still a significant size when loaded in as a blocklist on the DNS server, but it is a lot more reasonable and manageable than hundreds of millions or even billions of names.

New Domain Considerations

There is a side effect to this new domain approach: When a new legitimate domain comes online, it needs to wait a while before it is removed from the blocklist and becomes resolvable by others on the Internet. While more than 90 percent of new domains are likely malicious, the 10 percent that are legitimate could be important to businesses, so it is not feasible to hold them for extended periods of time before they can be considered safe.

So, let's ask a few business-centric questions: Does your business depend on being able to access new domain names that just appeared on the Internet? What is the cost of waiting a few days or more before your business can access the new domain? If you are the new domain owner, what is the cost to you to have to wait a few days before others can resolve your new domain? Does the new domain need to work immediately?[33]

For the majority of businesses, the answers to these questions are straightforward: Most can afford to wait a few days (or longer) before accessing a brand-new domain, with no significant downside or cost. This delay is generally perceived to be a small price to pay for the added security it provides. Of course, there will be exceptions, but those exceptions can be dealt with manually, such as RPZ's ability to explicitly allow resolution of a domain or IP address using an allow list.

Silver Lining

From a security perspective, the presence of all these domain names has a silver lining: They are almost always unique. In other words, two different malware destinations will rarely have the same domain name. A study[34] conducted in 2016 found that in over 160 million DGA names, there were only 17 instances where both shared the same name. That is not a typo. It is not 17 percent of 160 million; it's 17 names.

This trait gives us our silver lining when dealing with massive numbers of maliciously generated domain names: If we know the domain name the client queries, we know with high certainty what malware has infected the client. The bottom line is this: *By monitoring DNS traffic, we can identify which end clients are compromised and which malware family they are infected with.*

---

32 Source: Infoblox RPZ feed data

33 There will be exceptions, but you can manually configure RPZ to use explicit allow list or bypass.

34 http://www.covert.io/getting-started-with-dga-research/

Summary: Domain Generation Algorithms

Things may look bleak when you learn that threat actors use DGAs to generate millions of names, making techniques like RPZ less effective. However, the unique nature of malicious domain names means that by observing the DNS queries, we can accurately pinpoint which malware is active on the network. We can also increase the effectiveness of our defenses by employing the "guilty until proven innocent" approach, such as blocking new domains by default with techniques like NRD/NOD.

CHAPTER 5

DNS Tunneling

DNS as a Transport Mechanism, Part I: Two-Way Communications

How are attackers using the DNS protocol as a transport mechanism? In this chapter, we look at how an attacker establishes two-way communication with a DNS server, also known as DNS tunneling. Chapter 6, Part II, covers one-way communications used for data exfiltration.

As a security professional, you are already familiar with the concept of tunneling, sending data encapsulated within another protocol. In fact, this is how a VPN (virtual private network) works. DNS tunneling works the same way. It allows someone to send and receive data over the DNS protocol. If you were to inspect these DNS tunneling messages, they often show up as gibberish DNS lookups and responses. It is important to note that not all DNS tunneling is malicious. There are numerous legitimate applications that rely on DNS tunneling, including music streaming services, anti-virus software, content delivery network providers, and even mobile banking apps. One thing they all have in common is that they rely on DNS to carry data the protocol was not originally designed to carry. This is the reason that some DNS administrators consider all DNS tunneling to be a form of DNS abuse.

Below are some examples of legitimate applications or services using DNS tunneling:

- `g63uar2ejiq5tlrkg3zezf2fkemc6pi88tz.er.spotify.com.`
- `a-0.19-b3000081.010083.15e0.1d99.36d4.210.0.ic7arfsqqzf694fs8zf8nz2t9b.avts.mcafee.com.`
- `jp8zy7i7vawluximoxrko1p2tn58gj0fjjj2g.p.03.s.sophosxl.net.`

For example, Spotify uses DNS to figure out which server a specific song should be streamed from. Anti-virus vendors such as McAfee and Sophos send detected file hashes via a DNS tunnel to see if a particular virus pattern has been fingerprinted. You can't tell any of this by looking at these seemingly random domain names. That's the purpose of encoding the data when DNS tunneling—the data being sent is obscured from scrutiny. Whoever is doing the tunneling gets to decide what data is sent and how it is sent. It could be a new anti-virus definition being retrieved, a funny cat picture, or a malware payload. As security practitioners, we won't know until we have spent a lot of energy reverse-engineering the process.

Figure 14 shows how tunneling works in an attack scenario, but the principles are the same regardless of whether tunneling is benign or malicious. If this doesn't make sense yet, or if the long domain names are confusing, read this chapter and the next chapter, which provides Part II of the full explanation.

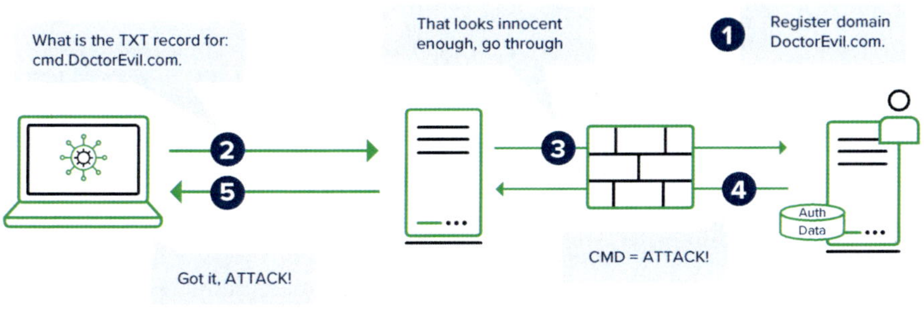

Figure 14: DNS tunneling

1. Attacker registered the domain `DoctorEvil.com`.
2. Compromised client queries for the TXT record of `cmd.DoctorEvil.com`.
3. DNS servers and network security devices allow this query to go to the attacker-controlled domain
4. The malicious command is incorporated into the DNS response
5. Compromised client receives the instruction

How Much Data Can DNS Carry?

You might be thinking, isn't DNS a lightweight protocol? How much data can attackers realistically carry over DNS? To answer that question, we must take a deeper dive into DNS resource records. Resource records constitute DNS data, and there are many different types of data, depending on the resource record type used. Below are six commonly seen DNS resource record types and one uncommonly seen type with a short description for each. The focus here is on how much data each record type can theoretically carry if we were to use it as a transport for DNS tunneling.

Keep in mind that when it comes to malware, attackers need to sneak payloads past security detectors. To do that, they need to spread the payload across a multitude of individual resource records. Depending on the payload, attackers could choose different record types to send different aspects of the payload.

The maximum data size ranges from a low of 4 bytes to a high of 259 bytes (excluding the rarely seen record type known as NULL). We'll learn shortly how attackers get around these size limitations when delivering payloads and how they use encoding to hide their presence. As you read the following sections, pay attention to both the domain name labels (left side) and the record-specific data (right side). There are other fields in the resource record, but they are beyond the scope of our discussion here, so we will not display them in the examples below.[35] The domain name reflects what the query may look like, and the record-specific data gives you an idea what the response is. For our purpose here, we will refer to the record-specific data field as the payload, because that is where the data being tunneled is placed.

A

The A (address) record type (generally called an "A record") maps a domain name to one or more IPv4 addresses. A records are the most commonly seen DNS record type today. They carry a maximum payload of 4 bytes. Example of an A record: `www.example.com. → 93.184.216.34`

AAAA

Pronounced "quad-A," the AAAA (IPv6 address) record type maps a domain name to one or more IPv6 addresses. It carries a maximum payload of 16 bytes. Example of a AAAA record:
`www.example.com. → 2606:2800:220:1:248:1893:25c8:1946`

CNAME

The CNAME (canonical name) record type links one domain name to another domain name. The full domain name can hold up to 255 bytes, but typically there are several labels (separated by the dot character). So, the more reasonable assumption is that it carries about 253 bytes (minus two or three dots). Example of a CNAME record:
`web.example.com. → www.example.net.`

MX

MX (mail exchange) records list the mail server for the domain. It contains an integer between 0 and 255 for the preference, as well as a domain name for the mail server. This yields 2 bytes for the preference, plus ~253 bytes for the domain name. Example of an MX record:
`example.com. → 10 mail1.example.com.`

SRV

SRV (service) records supply information about where to find network services such as LDAP. SRV records are typically intended for internal networks (e.g., they are heavily used with Active Directory). The SRV record includes three integers of 2 bytes each for various configurations, plus a domain name, yielding 6 bytes for the integers and ~253 bytes for the domain name, for a total of ~259 bytes.

---

35 Those other fields are TTL (Time to Live), class (almost always IN for Internet), and record type (A, AAAA, TXT, etc.). As an example, this is how an A record for `www.example.com.` would look if we were to display all fields: `www.example.com.    86400    IN    A    127.0.0.1`

Example of an SRV record:[36]

```
_kerberos._udp.example.local. → 0 100 464 dc01.example.
local.
```

TXT

TXT (text) records allow the storage of any generic text, and while each TXT record can only hold 255 bytes of data, many DNS servers allow the chaining of multiple TXT records together to send larger data sets, with no standardization on how many TXT records can be chained together. So, in theory, if the server implementation supports it, you can chain as many TXT records as you want to send a large amount of data. But with large payloads, you are likely to run into network limits[37] that may interfere with delivery of the large payload. Here are a few examples of TXT records:

```
example.com. 300    IN TXT "8cd468d7d5994fcc9d350683a8cb07a1"
example.com. 10     IN TXT "v=spf1 include:spf.example.com
~all"
example.com. 10     IN TXT "domain-verification=22r51cu0ab0sw5
36tlds4h"
```

NULL

Unlike the six types listed above, the little-known NULL record type was originally reserved for experimental purposes; it can hold up to 4,096 bytes of data. Modern DNS software almost never uses it for legitimate purposes. Because of its large size and its rarity, attackers who attempt to use it are at high risk of detection.

---

36 Keen observers may question the use of the underscore (_) characters used by SRV records. True, underscores are outside of the 63-allowed characters for hostnames. However, the underscore characters used here by SRV records are denoting services (such as LDAP or Kerberos), not hosts, thus they don't strictly follow the 63-character set for hostnames.

37 These limits may include but are not limited to: MTU (Maximum Transmission Unit), MSS (Maximum Segment Size), and IP Fragmentation.

Table 1 summarizes some of the most common DNS record types and the maximum bytes they can carry:

Record Type	Description	Maximum Payload
A	Maps a domain name to an IPv4 address	4 bytes
AAAA	Maps a domain name to an IPv6 address	16 bytes
CNAME	Links one name to another name	~253 bytes
MX	Lists a mail server for the domain	~255 bytes
SRV	Lists a network service location	~259 bytes
TXT	Holds generic text, can be chained	255 bytes and more
NULL	Experimental and rarely used	4,096 bytes

Table 1: Common DNS record types

You'll find information about which of these record types attackers use most often in the "How Attackers Avoid Detection" section later in this chapter.

Size Matters

Why are we discussing message size? Because when it comes to using DNS to transport malware and other payloads, size matters. Most DNS communications by design occur over UDP, which traditionally imposed a size limit of 512 bytes. With protocol extensions released in the year 1999, the limit was increased to 4,096 bytes.[38] Larger messages can also be sent via TCP, which the DNS protocol supports (TCP port 53). However, large network messages are at a greater risk of not reaching the destination. For example, legacy firewalls may still assume DNS messages are limited to 512 bytes in size.

The size limits for DNS messages are a constraint to a threat actor, who is probably trying to send additional malware components to an already-compromised device. An impatient attacker may want to transmit large packets to speed up communication—say, 4,096 bytes per message. However, large messages are less likely to reach the destination in one piece. To ensure delivery, smaller messages (shorter than 512 bytes) are much better. For this reason,

38 EDNS0 (Extension Mechanisms for DNS) allows DNS to transmit up to 4,096 bytes as a theoretical limit. The most recent community consensus is 1,220 bytes, based on practical experience, to avoid fragmentation. See DNS Flag Day https://dnsflagday.net/2020/

attackers who are trying to ensure the delivery of their malicious payloads tend to use hundreds or thousands of small packets, rather than one large transmission.

Encode, Encrypt, Oh My!

Size limitations are just one of several factors threat actors must consider when trying to use DNS tunneling for malware delivery. Encoding methods and encryption are two more. While the DNS protocol technically allows any character[39] (other than the dot character) to be included in a domain name label, common and acceptable encoding practice on the Internet is to use the same character set allowed for use when specifying the name of an individual host (i.e., its hostname, which is the left-most label in the FQDN).[40] The permitted hostname character set consists of:

- Alphabetical characters (A to Z, case insensitive)
- Numerals (0–9)
- Hyphen (-)

If they want to avoid detection, threat actors attempting to use a DNS tunnel are limited to these same characters. They could use other characters—even binary encoding would work. But that approach would be more likely to attract attention. It also risks having data stripped by other applications en route before reaching intended victims.

This character set limitation leaves the attacker with few choices when it comes to encoding data. Standard Base64 requires case sensitivity, and three additional characters that are not in the set used for hostnames (the + and / characters, plus = for padding). This means, if the attacker needs to send data over DNS using the hostname portion of the query, he or she must use an encoding standard with a smaller character set, such as Base32, or go with some customized encoding.

When it comes to using a portion beyond the hostname, however (as we covered in the "Hidden Path" discussion in Chapter 3), an attacker has more options available. While the hostname is limited to a smaller character set (alpha-numeric and the hyphen), the rest of the domain name is not subject to this limitation. Thus, threat actors can use a "legal" hostname, followed by the encoded data using characters such as /, +, and = in the other parts of the FQDN. Such encoding could be used as long as a compliant tag is added before the encoded data and separated by a . character. As an example, say the attacker has a bit of Base64 encoded data (`RE5TIEluc2VjdXJpdHk=`) to send. That data couldn't be sent in

---

39 See RFC 2181 section 11.

40 For certain record types that designate protocol services (e.g., LDAP in a SRV record), the underscore is commonly used; however, it is not a standard character allowed in Internet hostnames. Example: `_ldap._tcp.ds.example.com`.

a query for `RE5TIEluc2VjdXJpdHk=.example.com.` because the = would be outside the supported character set for hostnames. However, if we tacked on a header name before the data (e.g., `a1`), it could validly be carried through a DNS query (e.g., `a1.RE5TIEluc2VjdXJpdHk=.example.com.`).

Either way, character limitations for the host portion of a domain name can mean larger payloads are often necessary when messages are encoded using a smaller character set. The original data might be just 100 bytes, but by the time it is encoded with Base32, it might be 200 bytes. Depending on the encoding and encryption used, the attacker might be looking at transmitting 300 bytes of message just to send 100 bytes of actual data.

This method must seem horribly inefficient and tedious to the threat actor. However, what the attacker gains is a viable channel (DNS) for hiding their activity. In a large majority of networks, DNS traffic is scrutinized minimally or not at all. However, the size implications of DNS tunneling are good news for security professionals who are trying to detect and stop this activity. It means threat actors are very unlikely to transmit just a handful of DNS messages and be done. They are likely transmitting hundreds, if not thousands of messages, to get their communication across. This gives security professionals and researchers more time and a fighting chance to detect and stop the attack in progress.

How Attackers Avoid Detection

As we've just learned, if you are an attacker intent on successfully exploiting DNS tunneling, you must navigate several competing restrictions. On the one hand, the encoding schemes you use could result in larger message sizes, due to character limitations. On the other hand, you need to send data in small chunks, fewer than 512 bytes per message, to increase the likelihood of successful delivery.

But wait! As a malicious actor, you also don't want to get caught. Blasting large DNS messages will get your data across but is more likely to be noticed. A clever attacker would want to make the communication look as innocent as possible. Choosing the right type of resource record can be another factor in decreasing the likelihood of detection. Thus, threat actors who want to avoid detection would avoid these record types:

- SRV: While this can carry about 259 bytes of data, transmitting SRV records over the Internet will likely be noticed because it is not a common resource record type for clients to query for Internet domain names.
- MX: Normal clients do not query MX records; typically, only mail servers (e.g., MTA) query for MX records.
- NULL: This rare record type carries the most data, but it is also easily spotted because almost no application or clients use this resource record type.

To stay hidden, threat actors' best bets are the most common record types: A, AAAA, CNAME, and TXT.

InvisiMole and DNS Forwarding—a Cautionary Tale

As we've just seen, threat actors have a range of options for ensuring that their harmful malware payloads reach their destinations undetected using DNS tunneling. It might be tempting to think that organizations are safe from DNS tunneling as long as their networks are not directly connected to the Internet, as is often the case with networks that are highly segmented for security purposes. The InvisiMole example reveals the (security) holes in that thinking. Let's get to it.

Case Study: InvisiMole

```
Classification: Spyware, Malware
Active: Since 2013, observed in 2018
Specialty: Cyber espionage
```

The InvisiMole malware/spyware was first observed in 2018, but it has been in operation since at least 2013. This malware resurfaced in 2019 with updated DNS tunneling capabilities. The malware/spyware targets high-profile organizations in the military sector and diplomatic missions in Eastern Europe. It can record from victims' webcams and microphones, track their geolocation, and collect recently accessed documents, including photos from mobile phones connected to the infected computers.

DNS Forwarding—A Potential Hole in Your Layered Security Architecture

One notable DNS behavior associated with InvisiMole is its ability to tunnel data and communicate with C2 servers even from "secure" networks with no direct access to the Internet. How is that possible, you say? You can thank DNS forwarding. DNS forwarding is a common configuration where one DNS server forwards queries to another DNS server for resolution rather than performing recursion itself.

In this case, the DNS server that is having the queries forwarded to it is connected to the Internet. Figure 15 illustrates the spyware's communication path. A client device already infected with the spyware (e.g., a laptop infected via phishing) gathers sensitive data for exfiltration and connects to the "secure" network, which has no direct Internet access. The infected client then sends DNS queries to the DNS server A, which forwards the query to DNS server B, until it reaches the Internet and the attacker-controlled DNS server. The replies from the C2 servers follow the same path back to the infected client to retrieve the stolen data.

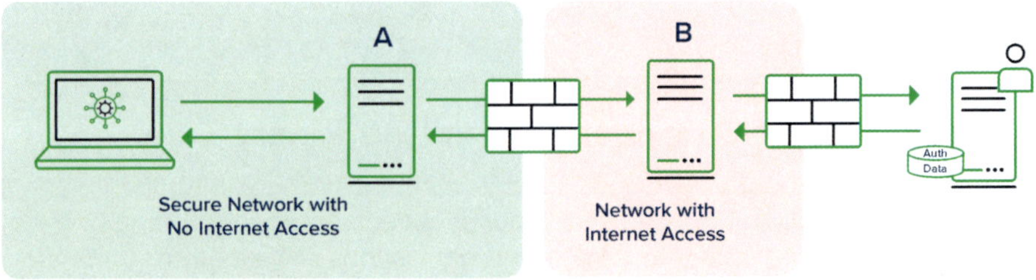

Auth Data = Authoritative Data

Figure 15: InvisiMole and DNS Forwarding

Summary: DNS Tunneling

Some people argue that DNS tunneling is an abuse of the protocol, no matter if it's done by an Eastern European malware author to command his or her global botnet or by your trusted financial institution to perform online mobile banking. The very act of DNS tunneling is transmitting arbitrary data—meaning data that was never intended to be delivered by the protocol—usually obscured by some type of encoding.

What does this situation mean for the security administrator? It means that an enterprise should not block all uses of DNS tunneling because there are valid applications and services that rely on DNS tunneling. As security professionals, we need an intelligent solution that can distinguish between legitimate and malicious DNS tunneling attempts; we also need to categorize these tunneling attempts. We need to allow the passage of benign and harmless communications while stopping or filtering malicious ones. In Chapter 6, we offer some criteria to help you to detect DNS tunneling in progress.

Luckily, due to the nature of the DNS protocol, attempts to encode arbitrary data via this attack vector will require multiple messages. In other words, compromised devices need to exchange multiple messages—sometimes ranging in the thousands—with C2 servers. If your security solution can detect this communication early in its tracks and stop it, you will likely stop the C2 servers from transmitting the malicious payload (such as additional ransomware components) to the compromised client.

CHAPTER 6

Data Exfiltration

DNS as a Transport Mechanism, Part II: One-Way Communications

You may look at the tunneling from Part I and think: Well, I can look at the responses, and if the responses contain malicious information, I simply won't pass them on to the client. Problem solved, right? After all, isn't that what RPZ is good for—blocking in-bound responses containing malicious information? In this chapter, we examine what happens when data theft happens from the inside out.

Data Exfiltration Over DNS

Unfortunately, even letting information leave your front door can be a problem in and of itself. Imagine if you have a tunnel from inside your house to the outside, where items only move outbound, not bi-directionally. What do you think the one-way tunnel is being used for? It is a perfect way for thieves to remove items from your house. That's data exfiltration over DNS. Traditional defenses rely on examining and blocking responses from the outside in. This approach is ineffective against the one-way communication of data exfiltration that flows from the inside out.

Figure 16 shows an example of DNS data exfiltration, combining the knowledge we have learned in previous chapters about tunneling and DGA:

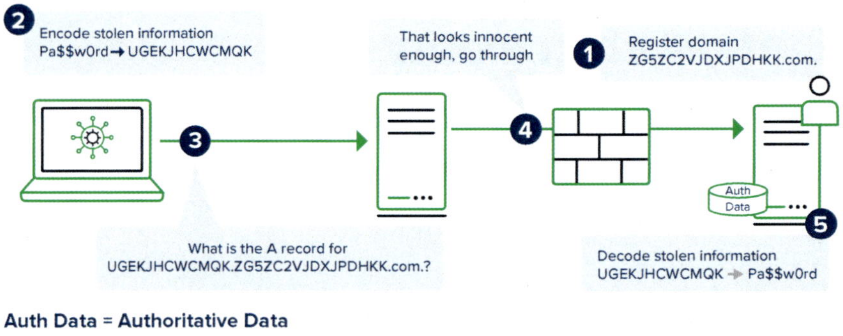

Figure 16: Data exfiltration over DNS

1. Attacker registers a random domain `ZG5ZC2VJDXJPDHKK.com`.
2. The infected device on premises encodes the stolen information. In this case, the string "Pa$$w0rd" becomes the encoded string "UGEKJHCWCMQK"
3. The encoded information is then sent to the attacker via DNS, by querying for the long, gibberish domain name `UGEKJHCWCMQK.ZG5ZC2VJDXJPDHKK.com`.
4. The DNS query is sent outbound
5. Attacker receives the query and decodes the stolen information. No reply is needed.

The dark genius of this technique is that it is one-way communication. For the method to work, the attacker does not need to send a response at all. A typical DNS server would send back an NXDOMAIN response because this random domain name probably doesn't exist. To avoid suspicion, the attacker's DNS server could send back some random IP address as its response. Or it could not respond at all and let this query time out.

The point is, as soon as the query is allowed outbound, the threat actor has succeeded in stealing your data. The threat is the query in this case, not the response.

Exfiltration and Zero Day Threats

Where data exfiltration is concerned, you might think: My DNS server and/or firewall is blocking with RPZ based on domain reputation (discussed in Chapter 3). So, if the attacker is using a known malicious domain, this communication would never happen in the first place because my DNS server or firewall would stop it. Right?

If only it were that simple. Unfortunately, attackers have learned to use newly minted domain names with good or no reputation in launching their attacks. Because these domains have no history of malicious activities associated with them, they are essentially zero-day threats. They won't be flagged by RPZs or security devices, even those that have advanced DNS threat intelligence built in. The malicious nature of these domains is not yet known, so their associated DNS queries are allowed to go through. Days or weeks may pass before we discover that an innocent-looking domain is doing something nefarious. But by then, it's too late.

Complicating things still further, some attackers are buying up expired benign domains with good reputations (known as farming) and leveraging those to further avoid detection in their exfiltration schemes.

Case Study: AlinaPOS

```
Classification: Malware
Active: 2012 - Present
Specialty: Targets point-of-sale (POS) devices for credit
card theft
```

The AlinaPOS malware event is a good illustration of the way DNS queries can be used in exfiltration. At one time, a POS system referred to a cash register; in the 21st century, POS is now synonymous with credit card processors or devices that accept credit card payments. They can be purpose-built devices or general-purpose computers running POS software. These devices are high-value targets for malicious actors because by design POS systems contain transactional data, including credit card information.

The AlinaPOS and its variants exploit Windows-based POS systems. Once infected, AlinaPOS scrapes through the device's storage (and RAM, if present) for account information and credit or debit card numbers.

However, removing this stolen information is tricky because payment processing systems are usually tightly monitored and reside in highly restricted environments. To work around this restriction, the group behind AlinaPOS has shifted toward using DNS as its transport to avoid detection and blocking. AlinaPOS uses a few domain names that look innocent at first glance and could pass for the content delivery network (CDN) provider Akamai. These domains include:

- `analytics-akadns.com.`
- `akamai-analytics.com.`
- `akamai-information.com.`
- `akamai-technologies.com.`

AlinaPOS crafts a DNS query with the following information:

- The first six characters are a unique identifier for the victim device, for example, cfjXVi
- The next few characters provide a short description of the victim device, such as DONOVAN-PC:1
- Finally, it includes the command to send to the C2 server (or if later in the C2 communication, the path of the executable) and the stolen payment information

Figure 17 illustrates the earlier communications from AlinaPOS-infected devices to the C2 servers. It includes the decoded information. The first query decodes to "Install," the second decodes to "Start" and the third decodes to "Ping."

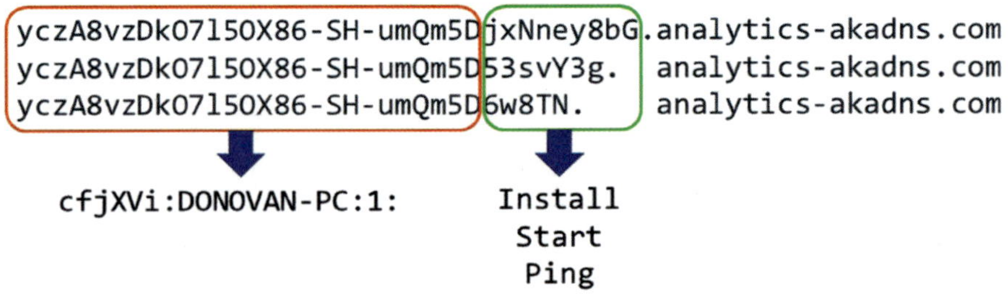

Figure 17: AlinaPOS DNS messages

Here is an example of an AlinaPOS query. The characters in bold constitute the exfiltrated data:

yczA8vzDkO7l5OX86-SH-umQm5CQ2sXZhM_
Sz5CQmZycmZ2dkpmTkpOemJ2cl5.
iYm5uYmpuampqampqbk5mam5qampqampKdnZqamg.analytics-akadns.
com.

This looks like gibberish to the rest of us, but when decoded by the attacker, this query string reveals the name of the device (DONOVAN-PC), the name of the binary that was executed (pos.exe), the payment information, and its expiration date. AlinaPOS is careful to send these queries as A record lookups because A is the most common type of DNS record requested. Combined with the domain names crafted to look like a legitimate service (Akamai), this exfiltration communication can be very difficult to detect, even for experienced security professionals. However, as we'll learn a bit later, there are steps we can take to detect this sort of malicious exfiltration.

Beyond Blocklists

Many commercially-available PDNS solutions today use RPZ (or something similar) as the sole mechanism to prevent resolution of known malicious domains. However, as we mentioned earlier in this chapter, RPZ is not likely to be effective blocking domains used in data exfiltration because those domains are often created just for that specific attack and have no prior reputation. Additionally, even if an RPZ could somehow capture one of these domains, an RPZ-enabled recursive server is often configured to not trigger until after resolution has occurred.[41] And as you now know, to succeed, data exfiltration does not require a DNS response. *The threat in this case is not in the response data, it's in the content of the outbound DNS query itself.* You can block and filter all the malicious responses, but once the query leaves your front door unchecked, you've lost the battle. We must do better than just looking at the target domain name. We need to analyze the outbound DNS query at a whole new level.

Deep Query Inspection

Inspecting for data exfiltration sounds more complex than it is. For example, we humans can intuitively spot suspicious patterns in DNS data. Look at the following four domain names of the sort you can easily extract from your DNS server query logs:

41 This is the default for many DNS servers for a good reason: Many domain names are linked to other domain names, so if you looked up good-name.com, it could link (CNAME) to evil-name.com, which we will not know unless we examine the response.

1. e6221.dscna.akamaiedge.net.
2. dualstack.osff2.map.fastly.net.
3. dualstack.apiproxy-website.prod-1.us-east-1.amazonaws.com.
4. yczA8vzDkO7l5OX86-SH-umQm5CQ2sXZhM_Sz5CQmZycmZ2dkpmTkp0emJ2cl5.iYm5uYmpuampqampqbk5mam5qampqampKdnZqamg.analytics-akadns.com.

Most people can tell that the last domain name in the list above is different from the previous three. But how? It's not just that the last one is longer. Other pattern recognition and linguistic analysis happened in our brain to arrive at that conclusion. To catch malicious outbound DNS queries in an automated fashion, we need to quantify these complex mental operations. For the following sections, we will break down these two names and compare them side by side (for the sake of space we will substitute the malicious string from #4 above with a slightly shorter one):

- dualstack.apiproxy-website.prod-1.us-east-1.amazonaws.com.
- yczA8vzDkO7l5OX86-SH-umQm5D6w8TN.analytics-akadns.com.

First, we will remove the domain names amazonaws.com. and analytics-akadns.com. because we assume technologies like RPZ will cover their classification. So really, we are comparing these two strings, String A (good) and String B (bad):

```
A. dualstack.apiproxy-website.prod-1.us-east-1
B. yczA8vzDkO7l5OX86-SH-umQm5D6w8TN
```

Detection Criteria

Here are four criteria to use when determining whether an outbound DNS query is most likely malicious or legitimate:

- Length of labels
- Lexical analysis
- Entropy of labels
- Frequency and number of total queries

Length of Labels

Normal DNS labels are typically not long, are usually something pronounceable, or are well-known abbreviations or acronyms. Table 2 presents the length of the labels in Strings A and B:

Label	Length
`dualstack`	9
`apiproxy-website`	16
`prod-1`	6
`us-east-1`	9
`yczA8vzDk07l50X86-SH-umQm5D6w8TN`	32

Table 2: Label length

Additionally, as we have established in earlier chapters, encoding information in DNS leads to long strings. Thus, many malicious actors attempt to use as many characters as allowed by the protocol per label (63) to get the stolen data out faster. More advanced malware may use shorter labels, say, 10 characters instead of 63, but we can use techniques such as lexical analysis to catch them.

Lexical Analysis

Lexical analysis is just a fancy way of saying: Normal DNS labels are generally words people can read, and malicious DNS labels are not. Even though few people read all of the DNS labels, people generally use recognizable words and expressions when creating legitimate DNS labels. Threat actors, on the other hand, have no interest in people reading these labels. In fact, it is in the threat actor's interest that no one ever reads these queries.

Let's take String A and break it down into individual labels:

- `dualstack`
- `apiproxy-website`
- `prod-1`
- `us-east-1`

Even if we may not fully understand the meaning of each label, we see short recognizable words such as "dual," "stack," and "east." Malicious DNS queries usually do not contain such logical and readable components. In fact, we cannot break down our String B example into any human readable components.

It is safe to say, if we don't see at least a few dictionary words in the DNS labels, it is likely a data exfiltration or tunneling attempt, especially if the characters appear to be completely random. Which leads us to the next analysis technique: entropy.

Entropy of Labels

Normal DNS labels are not random. DNS, after all, was invented because humans need domain names that are easy to remember, and a random collection of characters is hard to memorize.

Luckily, mathematicians already blessed the world with a way to quantify entropy, or the level of randomness. Figure 18 shows one of the common mathematical equations used to calculate the entropy of any given string, known as the Shannon Entropy.

$$H = -\sum p(x) \log p(x)$$

Figure 18: Shannon Entropy equation

When we run our strings of text through this equation, we get a very definitive score that tells us how “random” the string is. The higher the score, the more randomness or entropy is in the string. Table 3 lists the Shannon Entropy score for several strings:

String	Shannon Entropy
www	0
`mail`	2
`new-fp-shed.wg1.b`	3.61687
`dualstack.apiproxy-website.prod-1.us-east-1`	4.14674
`A34ilhu2398ysdvkjbwe`	4.22193
`yczA8vzDk07l50X86-SH-umQm5D6w8TN`	4.41391
`W.3217376901.IO.aHR0CHM6Ly91CGRhdGUu29vZ2xIYXBpcy5jb20V`	4.96282

Table 3: Shannon Entropy scores

To put things in perspective, the first four strings are part of real DNS domain names. The fifth string, `A34ilhu2398ysdvkjbwe`, with the score of 4.22193, is generated by randomly banging on the computer keyboard. The last two strings are part of DNS data exfiltration attacks. You may notice the score for the random keystrokes (4.22193) is not that different from the previous entry (dualstack). That is because identifying malicious patterns is only part of the puzzle. There are other factors to consider.

Frequency and Total Number of Queries

The last metric we can use to determine whether an outbound DNS query is suspect is to look at the number of unique queries associated with a given domain.

A normal DNS traffic pattern from a legitimate client to any given domain consists of a few, infrequent queries. There might be a surge of DNS lookups in the beginning when the client is trying to find the resource (e.g., a website), but after locating the resource, the client switches the bulk of the communication to other protocols (such as HTTPS) and DNS lookups stop. Compare this to a client controlled by the threat actor who is performing data exfiltration or DNS tunneling. A compromised client is likely sending hundreds, if not thousands, of DNS queries, all targeting the same domain name. Smarter attackers might spread this massive number of DNS lookups across several domains or across the span of a few hours or days to avoid detection, but the pattern should still be clear.

Detecting Anomalies

We have outlined some common characteristics of malicious outbound DNS queries used to exfiltrate data. These queries appear different from normal DNS queries, with longer labels, made up with a different mix of characters, and have a different frequency and total number of queries.

We can collect this information by reading DNS query logs. The retrieval of DNS query data can be enabled on most DNS server software. The logs can then be sent to a security information and event management (SIEM) system for collection, archiving, and analysis. More and more DNS vendors are adding capabilities for discovering anomalies. If your DNS vendor can detect DNS query anomalies, consider enabling that feature, in addition to analyzing log messages collected by a SIEM. Some security professionals may go even so far as to log the DNS responses, to examine both the query and response.

Below are some tips to help you spot malicious DNS patterns, once you have collected the DNS query (or response) data:

- *High volume of DNS queries for the same domain over a short duration of time:* This is also known as "rapid fire." Normal client behavior would not query for hundreds of subdomains within a few seconds, but a compromised host might do it, as it is trying to establish communication with C2 servers. For example, if you see a host querying for `abcxyz123.example.com.`, followed by `qweasd234.example.com.`, followed by `poilkj345.example.com.`, and many other names ending in `example.com.`, all within a short period of time, it is worth investigating.

- *Querying for long domain names with suspicious patterns:* As we have discussed earlier in this chapter, DNS is designed to be readable by humans, and even long domain names usually contain readable words. If you see a query for `dualstack.apiproxy-website.prod-1.us-east-1.example.com.`, although it is rather long, it is likely benign (as we have just discussed in the section above). But if you see a name such as `yczA8vzDkO7l5OX86-SH-umQm5CQ2sXZhM_Sz5CQmZycmZ2dkpmTkpOemJ2cl5.example.com.`, it should raise a red flag as it checks most of the boxes for a suspicious DNS pattern.
- *Querying for unusual record types:* While the record type alone may not be a reliable way to detect malicious DNS traffic, as we have discussed in Chapter 5, it can still be leveraged as one of the determining factors. For example, normal clients do not query for MX records; mail servers do. And while SRV record types are common on the local network (typically used for Active Directory), they are rarely used over the Internet. TXT records are common, but querying for and receiving numerous TXT records is less common. The biggest giveaway of all is the NULL record type: No common software or application uses it. Chances are anyone sending or receiving NULL records is doing so with a malicious intent.
- *Responses are consistent in size and contain suspicious patterns:* This is only available if you are monitoring DNS responses. Normal, organic DNS responses will vary in size. Bad actors who are trying to send a malicious payload to a compromised host via DNS will most likely maximize what can be sent in each response, usually resulting in consistent sizes. For example, if an attacker is trying to send a 25 KB (roughly 25,000 bytes) encoded malware payload to a compromised host over DNS, she may break down this malicious payload into 100 DNS messages, with each being exactly 250 bytes long (near the maximum size for TXT record). What we see over the wire is the client receiving 100 DNS responses, and each one is exactly 250 bytes in size. This pattern is unnatural in DNS and should arouse suspicion.
- *High volume of resolution failures or NXDOMAINS:* It is only possible to inspect for this if you are monitoring DNS responses. If you monitor DNS logs, you will likely find that there are many failures as part of normal DNS operations. However, that is not necessarily true for individual DNS clients. If a client consistently queries for a set of names that do not exist (which results in a response of NXDOMAIN), at best, it is misconfigured; at worst, the client is cycling through many names until it finds one that will allow it to contact a malicious C2 server, as described in Chapter 4.

DNS and SIEM

We can send DNS logs to a SIEM.[42] Various SIEM vendors have add-on features that can analyze DNS logs and detect security events, such as DNS tunneling. But beware, because some vendors' claims of protection against DNS tunneling or exfiltration are based on domain name reputation, not inspecting the outbound queries or responses in real time. These reputation-based defenses are less effective against zero-day or spear-phishing attacks.

Many SIEM vendors offer DNS analytic capabilities. Some, such as ArcSight's DNS Malware Analytics, are offered as a packaged add-on. Some offer community-sourced solutions, such as the numerous DNS apps available for Splunk. Some tools, such as RSA's NetWitness, allow security administrators to write custom rules looking for particular patterns.

Security professionals must understand the limitations of integrating DNS logs with a SIEM. DNS query logs may not contain IP address and port information or DNS transaction IDs, which are sometimes needed when attempting to detect patterns such as cache poisoning attempts. Query logs also do not typically contain response data and time-to-live (TTL) value, at least not by default for most DNS servers. That information resides instead in DNS response logs. If you find yourself in need of performing very deep level analysis, you may need to look for other collection tools, such as tcpdump or (net) flow.

It is also important to note that if you are relying on a SIEM to detect the presence of nefarious activity, you are in reactive mode and already a step behind your adversaries. Ideally, a proactive security approach would enable you to stop illegal DNS activity at the endpoint before it can even reach devices and SIEM tools. An effective way to adopt a proactive stance is to make DNS query inspection capabilities part of your DNS server. In such a configuration, the DNS server can both detect and block malicious DNS interactions in progress. Therefore, if you are building or selecting a PDNS solution, you should take deep query inspection and analysis capabilities into account.

Successful Attacks Get Better, Not Worse

道高一尺

魔高一丈

Chinese proverb, "Dao Gao Yi Chi, Mo Gao Yi Zhang," which loosely translates to: "When the angel grows by a foot, the devil grows by ten."

42 This could be, and often is, a tremendous amount of data. You should consider the cost and performance impact on your SIEM.

One of the ongoing challenges we face as security practitioners is that successful attacks have a tendency to grow stronger over time, not weaker.[43] With everything else on our plates day to day, we don't often have the luxury of keeping up with the latest trends in how DNS is being misused or abused. Unfortunately, the bad guys are paying close attention—not just to what they can do using DNS, but also to the tactics security teams are deploying to combat attacks. Threat actors continue to find creative ways to exploit DNS to stay ahead of those tactics, as the following case study shows.

Case Study: SUNBURST

```
Classification: Trojan, Malware
AKA: Backdoor.Sunburst, SolarWinds Sunburst
Active: Since 2020
Specialties: Advanced persistent threat, supply chain attack
```

In December 2020, major news stories reported the SolarWinds supply chain attack. The malware SUNBURST redefined our understanding of malware and brought the phrase "supply chain attack" into everyday vocabulary. This attack is notable because it managed to elude nearly all security products. Let's look at the role DNS played in this malware.

Once installed, the SUNBURST trojan/malware encodes the victim computer's information into a string and uses that string as part of the DNS lookup sent as part of its C2 communication. Figure 19 illustrates at a high level how SUNBURST C2 communication over DNS works:

43 Many security professionals assume this is a quote of Bruce Schneier, but in fact, is a quote from NSA: "Attacks always get better; they never get worse." https://www.schneier.com/blog/archives/2011/08/new_attack_on_a_1.html

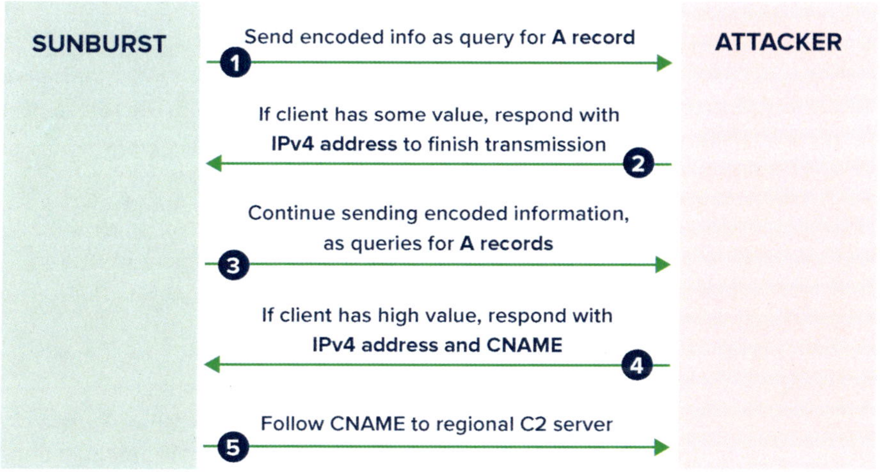

Figure 19: SUNBURST DNS communication

1. Infected client sends information about itself, including its Windows domain name, as part of the A record lookup. This query is sent to the attacker-controlled domain, `avsvmcloud.com`..
2. The attacker's DNS server processes the client's query, and if the query is of low value, the attacker does not respond; if the query appears to have some value, the DNS server responds with an IPv4 address.
3. Infected client continues to send information, including security products installed and what features are detected.
4. If the attacker's DNS server determines the client is of high value, it responds with an IPv4 address and a CNAME, pointing to a geographic location that is close to the compromised client.
5. Client performs subsequent lookups to follow the CNAME to a C2 server that is close to its geographic location and continues the communication over HTTP or HTTPS.

SUNBURST is a very sophisticated malware, employing a few new DNS-related techniques. We will focus on the three most pertinent techniques here:

1. Using IPv4 addresses as a means of communication
2. Using geolocation connections to avoid detection
3. Using domain names with good reputations (also known as domain name farming)

IPv4 Addresses as Means of Communication

The SUNBURST attacker is careful to send back IPv4 addresses that legitimate businesses own. The infected clients don't ever actually connect to any of the IPv4 addresses returned in steps 2 and 4 of Figure 19, so the attacker is free to use any legitimate block of addresses. These IPv4 addresses are used as a form of communication themselves. For example, if the client received an answer of IPv4 address in the 8.18.144.0/24 range (which belongs to Level 3 Communications), it will continue to send information over DNS about the infected device; if the client received an answer in the 24.140.0.0/15 block (which belongs to Microsoft), the client will restore disabled security product service registry keys and quit, essentially cleaning up the infection.

Geolocation

For clients that are of higher value, SUNBURST sends back a CNAME response (step 4 of Figure 19) to send them to the next set of servers for further exploitation. SUNBURST's DNS servers look at the client's source address and return answers that are closer to the client's geographic location. For example, an infected device in Los Angeles might receive a name that ends with `appsync-api.`**`us-west-1`**`.avsvmcloud.com.`, while a device in London might receive a name that ends with `appsync-api.`**`eu-west-1`**`.avsvmcloud.com.` (the geolocation labels are noted in bold).

The IP addresses for these corresponding names are also in the correct geographic location, which means that if your device in New York was infected, it likely made a connection to an IP address that is in the same region (the eastern United States). This level of evasion is unprecedented, and it makes detection very difficult.

Domain Name Reputation

SUNBURST uses `avsvmcloud.com.` as its main domain. While other domain names are used later during the CNAME response (step 5 in Figure 19), all initial DNS queries were sent to this one domain. This domain was first registered in February 2020 and maintained a good Internet reputation until at least October 2020. Other domain names used were similarly registered and used with care to ensure that they did not get flagged by reputation-watchers and appear on someone's blocklist.

This method tells us that the threat actor behind SUNBURST is well aware of the domain reputation process, and we can count on future malware to patiently stay dormant or for malicious actors to purposely cultivate domain names with good reputations (aka farming), only to strike when the time is right.

Summary: Data Exfiltration

By its nature, data exfiltration is mostly one-way communication. Traditional defenses that rely on blocking responses are thus ineffective. Many data exfiltration attempts hide their tracks among tens of thousands of domain names. While this may be effective to a certain degree, it is also costly to the attacker: Registration fees for all those domains can quickly add up. To keep the security industry guessing, more threat actors are using a combination of exfiltration, DGA, look-alike, and tunneling techniques.

A good security strategy should always be multi-faceted. As defenders, we cannot rely solely on blocklist technologies such as RPZ and NOD. Data exfiltration over DNS generates unusual and unique query patterns that are detectable. In this chapter, we presented four characteristics to evaluate: length of labels, lexical analysis, entropy, and frequency and number of queries. While these may not completely stop newer malware such as SUNBURST, they give the good guys a better chance of catching threat actors early. As security professionals, it is our duty to implement these detection techniques or deploy solutions that can do so.

CHAPTER 7

Cache Poisoning and DNSSEC

Insecurity in the DNS Protocol

So far, we have focused on how threat actors could misuse the DNS protocol to cause harm. Those attacks can be carried out without attacking the DNS servers themselves. In those scenarios, the DNS servers (usually recursive resolvers) are acting as the innocent messengers, sending and receiving on behalf of the threat actors. In this chapter, we discuss the messenger—the DNS recursive resolver—and focus on a specific type of attack that is especially useful to threat actors: cache poisoning.

While reading this chapter, please keep in mind that while the term "client" is used in the discussion that follows, both the cache poisoning attack and the protection techniques covered in this chapter apply only to DNS servers. Cache poisoning attacks target DNS recursive servers specifically. DNSSEC and DNS cookies are primarily used between DNS servers for added security. A brief discussion of cache corruption threats against client devices appears in Chapter 9.

What Is Cache Poisoning?

Recall that there are two main DNS server roles: recursive resolvers that will look for answers they do not already have and authoritative servers that provide answers to queries for their namespaces. As with many other Internet caching technologies, once the recursive resolver has received an answer, it stores the answer in its local cache for up to the TTL value. The next time a stub resolver (client device) asks the same query, instead of going all the way out to the Internet and asking other authoritative servers, the recursive resolver responds with the answer from its cache. The use of DNS cache can reduce Internet connection times, but that increased speed comes with a trade-off: A cache can provide an opening for abuse.

Cache poisoning is the general technique of tricking a recursive resolver into storing false information in its cache, usually with the intention to redirect clients to malicious destinations. For example, say the correct IPv4 address for the domain name `www.example.com.` is 10.1.1.1, but if I can somehow fool a recursive resolver into storing that address as 10.2.2.2 (or any IP address of my choosing) in its cache instead, that is a type of cache poisoning.

This technique is especially devastating when attackers target a recursive resolver that serves many users. For example, if the threat actor successfully poisoned the cache of an ISP's recursive resolver, thousands of ISP customers or users who rely on that recursive resolver are at risk. To make matters worse, cache poisoning is very difficult to detect. Many organizations that manage recursive resolvers do not even know they have been compromised. We will discuss later in this chapter the difficulties involved in detecting this specific type of attack.

A Brief Overview of Recursive Resolvers and Delegation

To better understand how malicious actors exploit recursive resolvers in cache poisoning attacks, it's helpful to know a few basics of a process known as DNS delegation. It should come as no surprise to learn that the DNS is essentially a vast, distributed database. DNS root servers do not store every single name in the world in a giant database. Not only would responses from the root servers be horribly slow, but it would also be extremely difficult to manage entries in such a large database. Instead, DNS root servers delegate responsibility for portions of the global namespace, such as `com.`, `net.`, and `org.`, to other servers. Those other servers each delegate further to the servers that control subdomains of their namespace—`example.com.`, `slashdot.net.`, and `isc.org.`, are a few examples. If you think of the DNS as many disjointed pieces of information, delegation is a directional sign that allows recursive resolvers to quickly locate specific resources from within that globally distributed database.

Let's use a short example to illustrate how recursive resolvers "chase down" delegations. Say Morpheus wants to know Neo's phone number. He knows the Oracle knows everything, so he calls the Oracle, who tells Morpheus to talk to Courtney. When Morpheus calls Courtney, she says she doesn't know Neo's number, but she knows that Satya works with Neo. So, Courtney refers Morpheus to Satya by sending him the name and number of Satya. Morpheus then calls up Satya, who may be able to provide Neo's number or, if not, refer Morpheus to yet another person, and this process continues until Morpheus finds the direct number to reach Neo. In DNS, these referrals of name and numbers are NS (nameserver) and glue records, respectively.

Essentially, when a recursive resolver (Morpheus) is looking for a domain name (Neo), it always starts by contacting the root servers (represented by the Oracle), which provides the answer (referral) pointing to the TLD (e.g., `com.`): "The name you are looking for is delegated to someone else (Courtney) to manage. Here is the *name* (NS record) and *address* (glue record) of the responsible party. That's who you should talk to next." The recursive resolver then presses on, querying the next name server (Courtney). If it gets another referral (e.g., Satya), it keeps asking other name servers until it receives a final answer to its question.

As the recursive resolver goes through this process, it adds each piece of data to its cache to help speed up the process for subsequent queries. For example, if it already has information for `com.` in its cache, it can talk to the `com.` servers directly, rather than having to go to the root servers first.

Once the recursive resolver gets the last bit of information needed to provide the answer to the original query (Neo's phone number), it checks that the information is valid before accepting the information and saving that data in its cache. This brings us to our next section, which discusses how recursive resolvers check the validity of the responses they receive.

Fooling Recursive Resolvers

You might be thinking: It must be hard to fool recursive resolvers. After all, DNS has been around for a long time; surely, we have figured out a way that recursive resolvers cannot be easily tricked into storing the wrong cache entry, haven't we? Let's take a quick look at the criteria a recursive resolver uses[44] to determine whether to accept or reject an answer:

- Source address and port
- Destination address and port
- Query ID (or transaction ID)

At first glance, this list may seem formidable. An attacker would need to fool the server in all these areas simultaneously. How hard is that to do? Let's break it down.

- *Source and destination addresses:* These are relatively easy to replace with bogus addresses because DNS query communication mostly takes place over UDP.[45] As you probably know, UDP requires no connection between the source and the destination before sending, so the sender (or attacker) is free to use any address they choose. In contrast, TCP's three-way connection handshake makes this type of address spoofing more difficult.
- *Source and destination ports:* The destination port offers zero protection from cache poisoning because it's fixed (port 53 for DNS). There's nothing for attackers to have to guess. The source port (also known as the temporary or ephemeral port[46]) is only slightly more challenging. While it's a potentially large number, the numbers are often reused and are relatively easy for attackers to figure out.

---

44 There are actually more criteria used, such as timing (how long since query), content (does the answer match the question asked?), and cookie. We purposely skipped these to focus on the particular protocol weakness that is being targeted by attackers.

45 Quick review if you don't remember the details of layer 4 technologies: UDP (User Datagram Protocol) is connectionless and easy to spoof; TCP (Transmission Control Protocol) is connection-oriented and harder to spoof.

46 Ephemeral port is just a fancy term to describe a semi-random port that is only used for the duration of the communication. Common DNS ephemeral port range is from 32768 to 65535.

- *Query ID (aka TXID):* Of these criteria, the query ID seems as if it should offer adequate protection from DNS spoofing. After all, it is a 16-bit integer that can potentially come in tens of thousands of combinations. That might have been enough in the 1980s, but not today. As we explain why, enjoy some pizza.

The Pizza Metaphor

To understand how attackers can carry out cache poisoning attacks, it's helpful to take a short snack break. Josh is hungry so he calls up his favorite pizza restaurant in town, Pizza Planet, and orders a large cheese pizza with extra anchovies. In about 20 minutes, a delivery person shows up at Josh's front door with the order. Being a security-minded person, and knowing that his nemesis Ross is out to get him, Josh asks the pizza delivery person: "Where are you from? What's the phone number of the pizza restaurant? What is my order number? What type of pizza is this?" The pizza delivery person answers all of these questions to Josh's satisfaction. Josh doesn't suspect a thing, pays the delivery person, and eats the pizza. But this pizza has been poisoned by Josh's enemy, Ross! Josh ends up in the hospital after eating the pizza, wondering: "How could this happen? I was so careful!"

Let's look at this from the attacker Ross's point of view as we go through each of the criteria we mention at the beginning of this discussion.

It is easy for Ross to figure out the addresses and phone numbers of Josh and the pizza restaurant (source and destination addresses): He can easily look them up in the phone book.[47] After Ross gets this information, he waits for Josh to place a call to order pizza (e.g., by tapping Josh's phone or simply knowing Josh's ordering habits). So now, Ross has almost all the information necessary to fool Josh. Ross knows the source and destination address. The only thing Ross doesn't have is the order number (transaction ID).

Ross is able to make a poisoned cheese pizza with extra anchovies that looks just like what Josh ordered. Ross also somehow guesses the order number correctly (you'll see that this is much easier than you think in the next section) and dispatches his minion with the tainted pizza. Josh asks all the questions that a DNS recursive resolver would ask, and because everything checks out, Josh accepts the poisonous delivery.

---

47 For our younger readers, a phone book is a database of name to phone number mappings printed on paper.

Lack of Entropy in DNS

The original designers of DNS created it at a time when network security wasn't much of a concern. And even if they did consider the security of the communication, they probably thought that it's nearly impossible to forge a DNS server response. The impostor would have to guess both the source/ephemeral port number (nearly 16 bits) and the query ID (16 bits). Combined, that adds up to almost 32 bits. That's the same number of bits as in an IPv4 address, which means the attacker would have to guess 1 in 4 billion! Surely, these odds are nearly impossible for anyone to exploit.

The Internet operated under this assumption for many years, until security researcher Dan Kaminsky discovered in 2008 that things are not as random as we previously thought. In other words, there is not enough entropy. The source/ephemeral port was not random enough, and many implementations often re-used the same port over and over. That made it very easy for an attacker to determine the exact port number or at least have a very educated guess.

So, you might think, take away the re-used port number, we still have 16 bits, or 65,535 possible numbers, for TXID. What are the chances of an attacker making a lucky guess to get 1 out of 65,535? And wouldn't it take a long time to guess? To answer this seemingly simple but in fact complex question, we go to the classic Birthday Problem.

The Birthday Problem

Seasoned security professionals know this classic probability problem brain teaser. A scammer hangs out at the local bar taking bets. The bet is: "I bet you $100 that two people at this bar share the same birthday." On the surface, there are 365 (or 366) days in a year, so naturally it feels like we need hundreds of people at the bar before we start to encounter any two people sharing the same birthday.

Will you take the bet? How many people do you think we need at the bar to have a 50 percent chance (or better) of two of them sharing the same birthday: 180? 90? 45? Being an experienced security professional, you likely already know the answer: 23. That's right, you only need 23 people to get to 50 percent chance of collision.

How many people would you need to achieve 90 percent certainty? We need 42 people. If we have 60 people, we reach 99 percent. So, if there are roughly 40 people at the bar, this is almost a sure win for the scammer: It is very likely (88 percent, in fact) that two people at the bar share the same birthday.

This may seem counterintuitive—how could the number be so low? One thing to consider is that for each additional person who enters the room, that person's birthday is now represented, reducing the set of potential birthdays without a collision. Figure 20 presents the formula for calculating the probability of collision (n = number of people, t = combination or number of days):

$$\textit{Probabilty of Collision} = 1 - \left(1 - \frac{1}{t}\right)^{\frac{n \times (n-1)}{2}}$$

Figure 20: Calculating the probability of collision

The Birthday Problem is a classic veridical paradox: something that looks intuitive and simple at first but is in fact complex. Put simply, as the number of people in the room increases, the likelihood of a birthday collision grows not linearly, but exponentially.

DNS Transaction ID Collision

We can apply the birthday problem principle to DNS transaction ID (TXID) collisions. Instead of 365 for the number of combinations, we use 65,535. And instead of the number of people in the room, we are now working with the number of queries, as in the number of queries the attacker needs to send with forged TXID.

We will save you the calculations. It takes roughly 300 packets for the attacker to achieve 50 percent collision rate, and 700 packets gets the attacker to 99.9 percent collision rate. In other words, if an attacker is trying to trick your recursive resolver into accepting his forged responses, probability theory says if he guesses 700 times, there's a 99.9 percent success rate that the attacker will guess the query ID correctly.

As our next case study demonstrates, threat actors that use cache poisoning have figured out how to improve those odds even more.

Case Study: The Kaminsky Attack

```
Classification: Cache poisoning
Observed: 2008
Tactics: Exploits DNS protocol vulnerability that affects
all major implementations of DNS servers, poisons cache on
recursive resolvers in as little as 10 seconds
```

As mentioned above, figuring out the source port in many cases was trivial, leaving the TXID as the only protection against a would-be attacker. The mathematics of probability dictates that if the attacker were to randomly guess a number in 65,535 query IDs, it would take 700 guesses to reach a 99.9 percent success rate. That's if the attacker plays fair. As we all know, attackers never play fair. Think like the threat actor again: What if I know what the victim is going to ask ahead of time? Figure 21 illustrates the cache poisoning technique that Dan Kaminsky used, which he disclosed in 2008. While studying the illustration below, keep in mind that the goal is to poison the domain name `www.example.com.`, but the attacker is using another name, `q0001.example.com.` to accomplish this mischief.

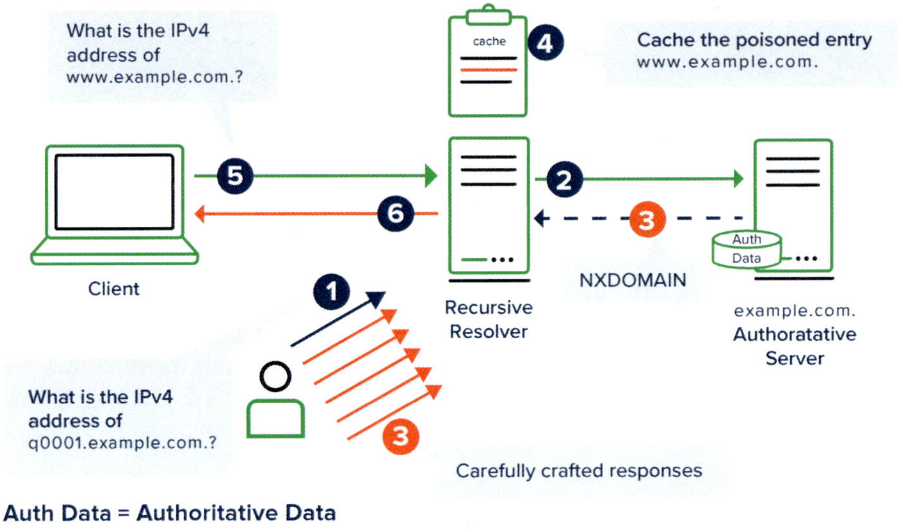

Figure 21: The Kaminsky attack

1. Attacker sends a normal query to the recursive resolver, asking for a name, such as `q0001.example.com.`, that does not exist.
2. Since this name `q0001.example.com.` does not actually exist, it is not likely to already be in the cache of the recursive resolver. The recursive resolver follows the normal name resolution process to find the authoritative name server for `example.com.` and queries it for this name. This step will result in the predictable NXDOMAIN response coming next.
3. Immediately after sending out the query for the non-existent domain name, the attacker starts blasting the recursive resolver with forged answers. The attacker has a head start before the real authoritative server can respond with NXDOMAIN. This head start allows the attacker to send hundreds of guesses of the TXID and play the probability game.

4. If the attacker is lucky in the previous step, one of the TXIDs matches what the recursive resolver is expecting, and the recursive resolver accepts the forged response and stores it in the local cache. The forged answer is specifically crafted so it poisons the names `q0001.example.com.` and `www.example.com.`; see details in the following section.
5. Unsuspecting client queries for `www.example.com.`, but this name is already stored in the cache on the recursive resolver.
6. Recursive resolver responds with the cached entry, which is the forged answer from the attacker.

How did Kaminsky poison the entry `www.example.com.`? In step 3, where the attacker is sending many forged responses, those responses are sending DNS information that tells the recursive resolver that the name `q0001.example.com.` is delegated to `www.example.com.`, which is on another DNS server. This is basically saying: "Oh, you want to find out about `q0001.example.com.`? That name is delegated to someone else to manage. You need to reach out to the server by the name of `www.example.com.`, IP address `10.11.12.13`." Figure 22 shows an example of this forged response; the TXID is 61718, the NS record is `www.example.com.`, and the glue record is `10.11.12.13`.

```
;; ->>HEADER<<- opcode: QUERY, status: NOERROR, id: 61718
;; flags: qr rd ra; QUERY: 1, ANSWER: 0, AUTHORITY: 1,ADDITIONAL: 1

;;; QUESTION SECTION:
;q0001.example.com.     IN A
;;; AUTHORITY SECTION
q0001.example.com.  2147483647 IN NS www.example.com.
;;; ADDITIONAL SECTION
www.example.com.  2147483647 IN A10.11.12.13
```

Figure 22: Forged response
Note: The transaction ID, NS record, and glue record are highlighted in red

Now we enter the probability game of guessing the TXID. Remember, the attacker most likely will win this guess, as the attacker can fire off 700 guesses to ensure a 99.9 percent success rate. Even if the attacker is ultimately unsuccessful (all 700 guesses are wrong), the victim recursive resolver gets a NXDOMAIN from the real authoritative server, and attacker can start over, this time using `q0002.example.com.`, another domain name.

If the attacker guesses the TXID correctly, which we have shown is very likely (700 guesses at 99.9 percent success rate), what happens? The recursive resolver accepts this forged response, it will be stored in the cache, and the attacker has successfully poisoned both the names `q0001.example.com.` (which was never the target) and `www.example.com.` (the real target).

Furthermore, the time-to-live (TTL) in the example is set to 2,147,483,647 (2 billion), the highest value allowed by the DNS specification.[48] That is roughly 68 years. This example illustrates that an attacker could set the TTL to something ridiculously long, hoping the DNS server never restarts and users will receive the poisoned answer for a long time to come.

To top all of this off, with this technique, Dan Kaminsky was able to poison the cache in about 10 seconds.[49]

To re-use our example from earlier in this chapter, if Agent Smith intends to trick Morpheus into believing Neo's number is his phone number, here is how he would try to accomplish this deceit. Agent Smith would contact Morpheus asking if he could provide the number for Bill1999 (a made-up name like q0001). Agent Smith knows Morpheus will first contact the Oracle because the name is made up, so it wouldn't be something Morpheus already knows (i.e., has in cache). Knowing that Morpheus is calling the Oracle, he starts bombarding Morpheus with responses posing to be the Oracle. Each forged response has a different TXID and says, "To find the number for Bill1999, contact Neo. Here is Neo's number." The number for Neo provided in the response is, in reality, Agent Smith's number. If Morpheus accepts one of those forged responses (the odds are in Agent Smith's favor as we have shown above), he will put that information into his cache and then call Agent Smith, who he thinks is Neo, to find out the number for Bill1999. The next time Morpheus wants to call Neo, he will look in his cache first, and call Agent Smith's number instead.

Detecting Cache Poisoning

Cache poisoning comes down to a game of chance—namely, guessing the query ID or TXID of the message. To increase the rate of success, threat actors send many forged responses with varying TXIDs, hoping to get one right. Smarter attackers would trigger the recursive resolver to query for the target domain name ahead of time, thus increasing the chance of success.

To detect possible cache poisoning in progress, security administrators or DNS administrators can look for unsolicited DNS responses with changing TXIDs. This is similar to our pizza example. If hundreds of pizzas showed up at your house after you've placed an order for one, and each pizza has a different order number, it should arouse suspicion. This detection is best performed on the DNS recursive resolver because it knows the definitive TXID. Thus, if numerous incorrect TXIDs arrive, the DNS recursive resolver can flag this arrival as a possible cache poisoning attempt.

48 RFC 1034 Section 3.6, TTL is a 32-bit signed integer in the units of seconds, giving us roughly 2 billion seconds.

49 There is an excellent and entertaining video explaining how this all happened, in Dan's own words: https://www.youtube.com/watch?v=B-v_wJIJUI4

As for the technique of triggering the recursive resolver to look up a domain name ahead of time, the best defense is to restrict who can perform recursive lookups. This approach is difficult if you operate an open recursive resolver where you accept recursive lookups from everyone. However, for most enterprises, there is no need to provide an open recursive resolver. Most enterprises have certain network blocks that are more trusted than others and can place rules or access control lists (ACLs) on the DNS recursive resolver to control this behavior. To wedge this into our pizza metaphor, this would be similar to making a rule at your house about who can use the phone to order pizza. If you simply allow anyone in the house to order pizza, and your door is wide open, the attacker could easily walk into your house and place an order from your phone.[50]

Although these two techniques are useful to prevent cache poisoning, how do you know if your cache has been poisoned? Unfortunately, there is no good answer here. This is what makes cache poisoning such a powerful tool for attackers. You could go through the names in the cache, compare them one by one by performing additional lookups against other DNS servers, but this method is resource intensive and generates a lot of false positives. DNS data is very dynamic, especially after adding factors like geolocation, load balancing, and content delivery networks (CDNs) into the mix. In short, your cache might say `www.example.com.` is 1.1.1.1, and when you look it up again, it might return 2.2.2.2, but that does not necessarily mean the cache has been poisoned. A crude method is to force your recursive resolvers to flush their caches more frequently. If you flushed your cache once every day, then even if the cache has been poisoned, it can only cause harm for at most one day. However, flushing the cache on a recurring basis would sacrifice some performance to gain greater security. Neither method is a good approach because we have not addressed the real problem.

The Fix, Circa 2008

In 2008, Dan Kaminsky proved that there was not enough entropy in the DNS protocol and server implementations. The protocol flaw was so fundamental, it was impossible to fix completely (except by moving to DNSSEC, which we will discuss later in this chapter). The lack of entropy resulted in a wave of DNS vendors producing patches, randomizing the source ports as well as improving randomization of the TXIDs.

What ended up being the fix (or "duct tape") was increasing the pool of entropy from 65,535 to 134 million[51] by implementing the source port randomization and improved TXID randomization, reducing the likelihood of collision. For the attacker, this means what used to take just 10 seconds to successfully execute

---

50 This is a very imperfect metaphor because the original attack technique is much more sophisticated, but hopefully our point gets across: Do not allow open recursion.

51 From 16-bits to 27-bits. http://unixwiz.net/techtips/iguide-kaminsky-dns-vuln.html

now takes hours, days, or weeks. Unfortunately, that's not completely eliminating the risk—in Dan Kaminsky's own words, "We just made the Internet less flammable."

The Internet has been content with this fix ever since. Many DNS operators put restrictions on their recursive resolvers to reduce the exposure to cache poisoning attempts. Few remember the risk of cache poisoning still exists. But for patient and persistent threat actors, as long as DNS remains the way it is, cache poisoning is still a possibility, as this next case study shows.

Case Study: SadDNS

```
Classification: Cache Poisoning
Observed: 2020
Tactics: Used advanced ICMP port knocking and fragmentation
to reduce entropy and successfully perform cache poisoning
in minutes
```

Many people think that since cache poisoning was "fixed" back in 2008, that it's no longer an active threat on the Internet today. They are wrong. First, the original Kaminsky-style attack never went away. Attackers are just taking longer to compromise caches. Second, victims of cache poisoning usually don't know they have fallen prey to malicious actors. Third, computational resources have increased exponentially. What used to take attackers weeks now takes days (or less). Finally, because this is a protocol-level flaw that could not be fully fixed (meaning it is still open today), attackers are continually improving their tactics to find more efficient ways to exploit it.

Disclosed late in the year 2020, Side channel AttackeD DNS (SadDNS) is a sophisticated attack that basically reduced the difficulty of collision by defeating the "fixes" implemented in 2008.[52] SadDNS is able to successfully poison the cache in several minutes. It does so with the following two techniques: (1) ICMP knocking to defeat port randomization and (2) fragmentation to avoid guessing TXIDs.

ICMP Knocking

Using clever ICMP messages and a divide-and-conquer technique, SadDNS can quickly narrow down which source or ephemeral port is open. Recall that when Dan Kaminsky discovered the protocol vulnerability, it was a matter of finding a collision in a pool of 65,535 numbers (16 bits). The patch from 2008 increased entropy to a total pool of 134 million (27 bits). SadDNS reduced the pool to 131,070

52 The techniques described here have been verified in 2022 to be effective: https://indico.dns-oarc.net/event/42/contributions/909/attachments/866/1568/Frag-DNS-OARC-2022.pdf

numbers (17 bits).[53] In other words, before the patch, it takes 10 seconds to poison the cache; after the patch, it takes days or weeks; with SadDNS, it takes a few minutes to breach the cache.

Fragmentation

SadDNS has another trick up its sleeve—to bypass any guessing altogether. In a normal DNS message, the UDP port and query ID information are in the beginning of the message, while the answers are near the end of the message. The attacker can cleverly split the UDP message in half at the right place, by lying about the MTU (maximum transmission unit) value, so that the message splits into two fragments as Figure 23 shows. Then, all the attacker needs to do is transmit the forged second fragment, skipping guessing the port number and TXID altogether.

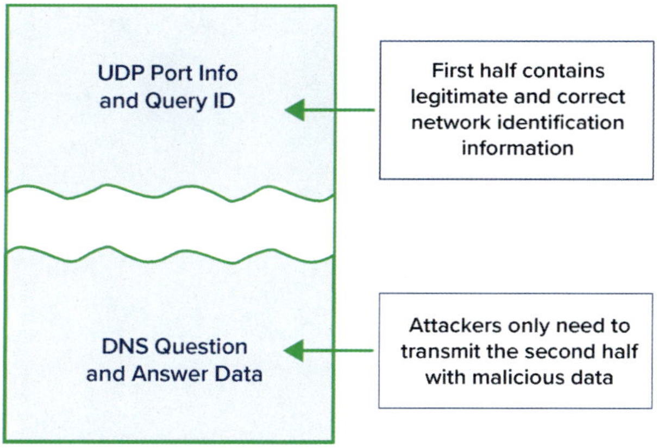

Figure 23: Fragmented DNS message

DNSSEC, the Real Fix for Cache Poisoning

Attacks like cache poisoning work because, as we mentioned earlier, DNS queries are transported over the connectionless UDP protocol, which lacks fundamental protection against malicious or forged answers. The DNS Security Extensions (DNSSEC) address this need, by adding digital signatures into DNS data so that each DNS response can be verified for integrity (i.e., the answer was not tampered with during transit) and authenticity (i.e., the answer came from the true source, not an impostor). In the ideal world, when DNSSEC is fully deployed, every single DNS answer can be validated and trusted, eliminating the threat of cache poisoning.

53 https://blog.cloudflare.com/sad-dns-explained/

To be clear, the term *DNS security* is NOT the same as *DNSSEC*, though the two terms are often confused. As shown in Figure 24, the term "DNS security" covers a broad range of security topics and measures related to DNS. The term "DNSSEC," on the other hand, is just one of those measures, one with a very specific meaning: It is a set of protocol update specifications.[54]

DNSSEC has two sides: the validation (recursive) side and the signing (authoritative) side. As you read the following sections, keep in mind that while DNSSEC uses cryptographic keys, the protocol provides *no privacy*. We will discuss a solution for the privacy gap in Chapter 8, "Encrypted DNS."

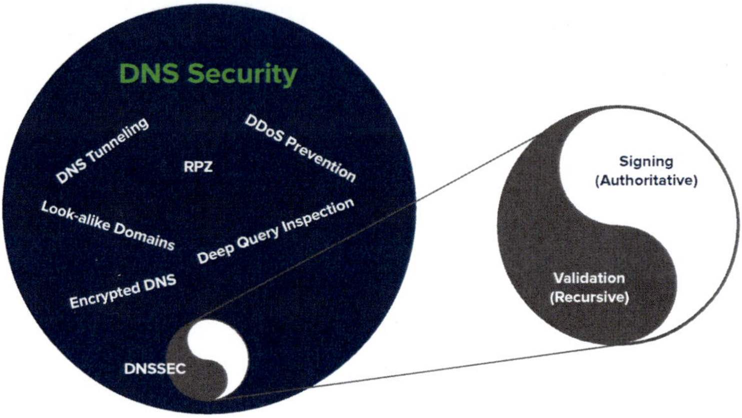

Figure 24: DNS security and DNSSEC

Over the years, DNSSEC has gone from a "nice-to-have" to a "must-have" for many organizations. The United States government requires all federal entities to implement DNSSEC.[55] Some European governments and municipalities such as the Netherlands[56] mandate DNSSEC for their domains. Rumors that standards such as PCI DSS and GDPR may soon require DNSSEC have been floating around for years. In addition, some security auditors have begun to add DNSSEC to their checklists.

54 DNSSEC specifications were originally described in three RFCs: RFC 4033, 4034, and 4035. There have been several updates since the initial specification.

55 This requirement has been in place since 2009, https://www.whitehouse.gov/wp-content/uploads/legacy_drupal_files/omb/memoranda/2008/m08-23.pdf

56 https://www.forumstandaardisatie.nl/sites/bfs/files/proceedings/FS-191009.5C2_Information_for%20MS_on_DANE_policy_NL_gov-ano.pdf

DNSSEC Features

DNSSEC adds four features previously unavailable in traditional, or insecure, DNS:

- *Authenticity, "I can prove who I am":* Also known as origin authentication. Using digital signatures and cryptographic keys, authoritative servers hosting DNSSEC-signed domains can respond with cryptographically signed data to prove the data's authenticity to others.
- *Data Integrity, "I can guarantee the content was unchanged during transit":* DNSSEC-enabled authoritative servers respond to queries with both plaintext answers and digital signatures. DNSSEC-enabled recursive resolvers (also known as validating resolvers) can compute the digital signature, thus verifying the plaintext answer was unchanged during transit.
- *Chain of Trust, "Other people can verify who I am":* DNSSEC takes advantage of the DNS hierarchical structure, allowing validating resolvers to verify the trust relationship between child domains (e.g., `example.com.`) and the parent domain (e.g., `com.`), until it reaches an explicitly configured trust anchor (usually the DNS root). This approach requires DNSSEC administrators to upload information to the next-level parent zone in order to connect to the chain of trust. They accomplish this task with a delegation signer (DS) record in the parent domain that contains a hash of the public key signing key (KSK) which is stored as a DNSKEY record in the child domain.[57]
- *Proof of Non-Existence, "When I say something does not exist, I can prove it":* With DNSSEC, when a recursive resolver is validating DNSSEC queries for a name that does not exist, the server can prove it definitively, beyond a simple NXDOMAIN response. This feature prevents scenarios where attackers cause denial of service (DoS) by spoofing DNS responses, such as lying to clients that `www.google.com.` does not exist.

DNSSEC High-Level Overview

DNSSEC is a protocol update. It provides extensions to both of the dual roles DNS servers play: recursive (validation) and authoritative (signing). You can deploy DNSSEC to each role independently—for example, to enable DNSSEC validation only for recursive lookups, but not for signing your authoritative zones.

Figure 25 shows at a (very) high level how DNSSEC works. As you study this illustration, keep in mind what needs to be in place before communications begin. In the center, a trust anchor has been installed on the validating resolver. It represents someone or something the validating resolver explicitly trusts. In DNSSEC, this is usually the root. On the right, large green arrows signify that the administrators of the child must upload authentication data to the parent.

---

57 There is also another key type, the Zone Signing Key (ZSK) that is also stored as a DNSKEY record type. We have purposedly skipped details about DNSSEC and new record types in order to focus on how DNSSEC works at a high level.

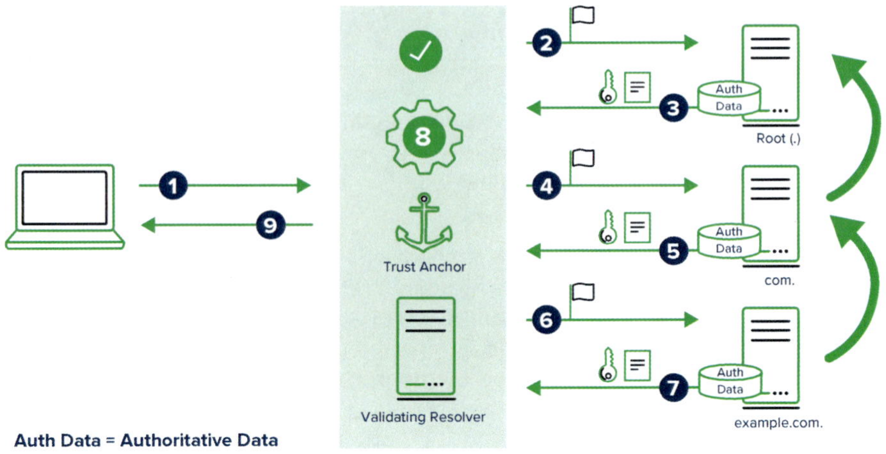

Figure 25: DNSSEC high-level overview

1. Client sends a regular (insecure) recursive DNS query to the validating resolver for something in the `example.com.` domain.
2. Validating resolver sends outgoing query with DNSSEC OK (DO) flag, signifying it wants to perform DNSSEC.
3. Root responds with plaintext answers and digital signatures, referring the validating resolver to its child, `com.`, to resolve something in the `example.com.` domain.
4. Validating resolver sends a query with DNSSEC OK (DO) flag to name server for `com.`, signifying it wants to perform DNSSEC.
5. Name server for `com.` responds with plaintext answers and digital signatures, referring to `example.com.`, its child.
6. Validating resolver sends query with DNSSEC OK (DO) flag to name server for `example.com.`, signifying it wants to perform DNSSEC.
7. Name server for `example.com.` responds with plaintext answers and digital signatures.
8. Validating resolver follows the chain of trust to perform validation work, which could involve going back to each name server at different levels to gather additional data, such as DNSKEY and DS records (if that information is not already in cache), until the validating resolver reaches a key that matches the configured trust anchor.
9. Validating resolver responds to client. There are three possible responses: secure, insecure, and bogus.

DNSSEC Responses to Clients

In the final step of Figure 25, the validating resolver responds to the client in three possible ways:[58]

- *Secure:* The target domain deployed DNSSEC and the validating resolver verified everything. The answer can be fully trusted. An additional authenticated data (AD) flag is set.
- *Insecure:* The target domain did not deploy DNSSEC and the validating resolver fell back to traditional/insecure DNS resolution. The answer is not validated.
- *Bogus:* The target domain deployed DNSSEC and during validation, the validating resolver detected mistakes. The domain is either misconfigured or compromised. Clients most likely just see the SERVFAIL response code.

It might be easier if we go back to the earlier pizza delivery metaphor. If we were to perform DNSSEC validation when trying to determine whether to accept the pizza, then these three responses mean:

- *Secure:* I was able to fully verify this pizza really came from Pizza Planet: I verified all the way up to the owner. You can eat this pizza without worry!
- *Insecure:* I was not able to verify one way or another if this pizza is safe. Consume at your own risk.
- *Bogus:* This pizza is definitely bad. Do not accept it!

DNSSEC Misconceptions, Clarified

Many people have false assumptions about DNSSEC, in part due to its generic-sounding name. In fact, DNSSEC is a very specific solution to solve a very specific problem (cache poisoning). We will try to debunk a few common DNSSEC misconceptions in this section.

Misconception: DNSSEC Provides Data Privacy

Perhaps the most common misconception about DNSSEC is that it provides privacy. It is easy to see where this comes from. We see words such as "cryptographic keys" and "security." However, DNSSEC is only using public key cryptography for authentication, not privacy. This means that anyone can still see the plaintext answer, along with the digital signature (RRSIG) used for validation. Figure 26 shows a sample DNSSEC lookup, with RRSIG highlighted. The query is for the name `www.example.com.`; the answer is 93.184.216.34. As you can see, even when DNSSEC is used, there is no data privacy.

---

58 Technically, there are four possible responses according to RFC 4034 section 5, the fourth type being indeterminate. However, most DNSSEC servers only send three types of responses.

```
dig @8.8.8.8 www.example.com. A +dnssec +multi

;; ->>HEADER<<- opcode: QUERY, status: NOERROR, id: 43941
;; flags: qr rd ra ad; QUERY: 1, ANSWER: 2, AUTHORITY: 0, ADDITIONAL: 1

;; ANSWER SECTION:
www.example.com.   20694 IN A 93.184.216.34
www.example.com.   20694 IN RRSIG A 8 3 86400 (
     20220108012601 20211217231841 31944 example.com.
     FtjRmXLGYH+5TJfzzfUfmqijGDmkKYBL/mno3cMvbEsL
     VANw6ENtM2juY+c3Xl1DKIQygu3ekrpzueWexnEKTUVD
     BBYwJgNgT0hIqnZ0PcrT7y+eBG9FwSiwsra+4yrUXbTB
     wZ6pd883KwGvNwOohHudTD3mt7SGYZ1d4Vbv+Lk= )
```

Figure 26: Authenticated DNSSEC response

The signature is highlighted in red, also note the "ad" in the flags section, representing that this is authenticated (i.e., validated) data.

Misconception: DNSSEC Needs Certificates or PKI

The phrase "public key cryptography" may automatically trigger the acronym PKI for many security professionals. The truth is public key infrastructure (PKI) is mainly needed for key distribution. Because DNS itself contains the publishing and distribution mechanism, it does not rely on any external PKI, such as the certificate authority (CA) system. In a typical DNSSEC deployment, keys are generated, kept, and published on the same set of DNS servers. You do not need to purchase certificates for DNSSEC.

Misconception: DNSSEC Protects Users From Malicious Domains

DNSSEC is a protocol that allows DNS servers to authenticate responses from another DNS server. However, DNSSEC does not prevent a threat actor from registering a domain name, signing it (i.e., to deploy DNSSEC on authoritative data), and using the domain for nefarious purposes.

Let's take a simplified example. Ross registers the domain `attckr.com.` and signs the zone data, thus deploying DNSSEC on the authoritative side. Josh enables DNSSEC validation on the recursive resolver, turning it into a validating resolver, thus deploying DNSSEC on the recursive side. Ross sets up the resource record for `ftp.attckr.com.` pointing to the IP address of a server hosting his malware. When Josh sends a query for `ftp.attckr.com.` to his recursive DNS server, the server will use DNSSEC to fully validate this name, authenticating that it really came from Ross. But DNSSEC does not, and cannot, know whether `ftp.attckr.com.` is a safe place to visit. If you want to filter malicious domains from DNS, use such technologies as RPZ because DNSSEC provides no filtering.

State of DNSSEC Adoption

As previously stated, DNSSEC is a specific solution to a specific problem: cache poisoning. But in order for DNSSEC to be effective, it must be more broadly adopted. Like DNSSEC itself, its adoption has two sides: recursive (validation) and authoritative (signing) and they are moving at different rates. As you read the following discussion, take note that the domains and regions with a currently high DNSSEC deployment rate are those where local policies and compliance regulations require it. Be aware that even if DNSSEC is not yet required in your region today, you may be just one compliance update away from having to add it to your organization's mandatory security check list.

The exact deployment status of DNSSEC is difficult to measure. By rough estimation, 22 percent[59] of the Internet's recursive resolvers, such as Google's public DNS service (8.8.8.8), perform DNSSEC validation when they look up domain names. The regional[60] breakdown paints a clearer picture. In Poland, validation is 50 percent, United States 40 percent, Australia 30 percent, and Japan 20 percent. On the authoritative (signing) side, deployment status differs greatly by top-level domain. Zones in `.gov.` are about 90 percent[61] signed, whereas zones in `.com.` are approximately 1 percent signed. DNSSEC is a complex topic, but there is no doubt where the world is headed. The DNSSEC section in this book is only a basic overview. To learn more, a good starting place is the Internet Society's DNSSEC page: https://www.internetsociety.org/deploy360/dnssec/.

Deploying DNSSEC

By now, you should have a good grip on the threat (cache poisoning) and the fix (DNSSEC), and you should plan on deploying DNSSEC for your environment. As mentioned earlier, DNSSEC comes in two flavors: validation (recursive) and signing (authoritative). Think of the validation as the web browser that supports HTTPS and the signing as the websites that offer HTTPS connections: You need both sides. For most enterprises, deploying DNSSEC validation can be as simple as changing the DNS forwarding path to a provider that supports it, such as an ISP. Implementing DNSSEC signing is much more involved and requires careful planning and execution, as well as coordination with your domain registrar to implement.[62] Signing will also require changes and additions to ongoing operational processes for your DNS infrastructure.

---

59 22 percent data is from 2019, https://blog.apnic.net/2020/03/02/dnssec-validation-revisited/

60 Current validation data map https://stats.labs.apnic.net/dnssec

61 https://fedv6-deployment.antd.nist.gov/snap-all.html

62 The need for planning and coordination with the registrar does not end once the DNSSEC deployment is complete. DNSSEC keys are rolled over periodically, meaning that there needs to be planning, execution, and coordination on an ongoing basis. Some proposals and standards such as RFC 8078 aim to reduce these maintenance efforts but are not widely supported yet.

For security professionals who plan to work with their DNS infrastructure team to deploy both DNSSEC validation and signing, we suggest the following priorities:

1. *DNSSEC validation (recursive):* Because DNSSEC validation is the easier of the two to deploy, and can usually be done with minimal impact, you should have a short-term plan to enable it, turning your recursive resolvers into validating resolvers.
2. *DNSSEC signing external authoritative domains:* Work with your DNS infrastructure team to have a mid- to long-term strategy to enable DNSSEC to sign your externally accessible domains, wherever practical.[63]
3. *DNSSEC signing internal authoritative domains:* In short, don't sign internal private domains. More explanations follow.

Internal Zones and DNSSEC

The question invariably comes up, should organizations sign internal zones with DNSSEC? In short, *no*. DNSSEC for internal or private domains is very rare, has additional complexities, and most enterprises won't benefit from it. In fact, it's generally a bad idea. In the first place, with internal zones, there is no way to tie the internal (private) namespaces to the chain of trust anchored at the Internet root servers. In the second place, because in many cases dynamic updates are performed to internal domains, DNSSEC signing internal namespaces should be avoided, because each dynamic update would require the entire zone to be re-signed.

The truth is, there is little to gain from enabling DNSSEC internally in most environments because often the recursive and authoritative roles are collapsed onto the same internal DNS servers, making signing of internal zones completely pointless. Put another way, if DNSSEC were deployed internally, and the internal server is functioning both as authoritative and validating resolver (a common and practical deployment), the DNSSEC server essentially tells the client: "Trust me because I validated myself, and I declare myself trustworthy."

Our recommendations are: First, successfully deploy both recursive and external authoritative DNSSEC. Then assess whether signing your internal domains is absolutely necessary. In almost all cases, the answer is no.

63 There are instances where certain technologies, such as some global server load balancing or apex record flattening technologies, may make signing a domain difficult or impossible.

If your goal is to deploy DNSSEC in your environment, we have not provided nearly enough details in this book. It is not our intention to provide a comprehensive guide for DNSSEC deployment. If you want to learn more about deploying DNSSEC, the ISC DNSSEC Guide,[64] originally written by one of the authors of this book, is a good start.

While You Are Waiting for DNSSEC, Have Some Cookies

You might think to yourself: DNSSEC sounds hard! And you'd be half-right. Enabling DNSSEC on the recursive resolvers is straightforward; enabling it on the authoritative servers requires much more planning and coordination. Although the industry consensus is that we all need to move toward DNSSEC, no one knows how long it will take for the entire Internet to get there.

Fear not. There is another technology that provides some increased protection against cache poisoning and reflection/amplification attacks—DNS cookies. Like DNSSEC, DNS cookies come in two flavors, client cookies and server cookies. Unlike DNSSEC, you probably already use cookies today. You may just need to fine-tune your configuration. Cookies are both easier to understand and enable than DNSSEC (client cookies are generally already on by default and have been for years).

However, this simplicity comes at a cost. DNS cookies only protect you from off-path attacks—ones where bad actors are unable to see your DNS information. DNS cookies offer no protection for on-path attacks where the malicious actor is positioned between DNS servers and clients and hence has access to your network traffic. In addition, cookies only provide validation that the server an answer came from was the server queried on a hop-by-hop basis rather than end-to-end like DNSSEC provides. DNS cookies also provide no way to ensure that the data in the answer itself is authentic. Despite these limitations, DNS cookies still substantially reduce the risk of cache poisoning, as we'll see in a moment.

DNS server cookies are likely already a feature on your DNS servers, but you need to ensure that they are enabled. You can enable them without worrying about stub resolvers or clients (because those are enabled by default). You should work with your DNS team to review whether your current DNS servers support DNS server cookies (they likely do), and work on a plan to enable and require their use.

---

64 https://bind9.readthedocs.io/en/latest/dnssec-guide.html

DNS Client Cookies: When sending a query, the client (a stub resolver or a recursive resolver) attaches a client cookie (which is a 64-bit pseudo random string) to the query. When responding, the server sends back the client's cookie, along with the server's own cookie. When the client receives the answer to its query, it caches the server cookie and compares the client cookie returned from the server with its own. If the client cookie matches, the client knows that the answer came from the DNS server that the query was sent to, rather than a bogus answer from an impostor. Because the chance of a bogus answer being accepted by a recursive resolver using cookies is lower, the chance of a successful cache poisoning attack on a recursive server decreases significantly. The use of client cookies is enabled by default in most modern operating systems and DNS servers, so chances are you have no action to take in order to benefit from client cookies.

DNS Server Cookies: Server cookies are used a bit differently. They can help a server validate that a query is coming from a legitimate client (i.e., it is not spoofed), which helps mitigate the risk of reflection/amplification attacks leveraging that server. When a server receives a query with a server cookie, it compares that cookie to its own. If they match, then the server knows it came from a legitimate client and can allow it to bypass certain other security measures. If they do not match, the server can take further action to ensure that the client is legitimate, such as limiting the response size or instructing the client to retry its query via TCP.

When DNS client and server cookies are in use, both the client and server will have some level of confidence they are talking to each other and not an impostor. Let's revisit our pizza metaphor to compare how DNSSEC and DNS cookies work.

Pizza delivery via DNSSEC: With DNSSEC enabled, when Josh receives a pizza delivery from Pizza Planet, not only can he check the order number on the pizza, but he can also see that there is a lock on the pizza box, proving the pizza has not been opened during transport. Although unlike a cardboard pizza box, imagine the pizza is in a clear plastic container so everyone can see what Josh ordered—there is no privacy. Josh also checks the driver's license of the delivery person. Furthermore, Josh calls up the delivery person's boss, the boss's boss, and so on all the way until Josh reaches the owner, who is a personal friend that Josh trusts to validate that the pizza came from Pizza Planet. This is a simplification of how DNSSEC works, but even with this basic scenario, you can see that there are many parties involved: *in order for it to work, everyone in the chain must support DNSSEC.*

Pizza delivery via DNS cookies: Things are a little easier with DNS cookies. When Josh calls the restaurant, he tells the person taking the order a secret phrase say, "yellow55frog." This secret phrase represents the DNS client cookie. When the pizza is delivered, a note comes with the pizza that says "yellow55frog-blue87tree." This combined string is the DNS server cookie. Note it begins with the client cookie, which Josh uses to validate it came from Pizza Planet. When

Josh needs to place any more pizza orders, he could use the server cookie, which allows Pizza Planet to know that the order came from Josh rather than an impostor.

Relying on a server cookie is effective only to a certain extent, though. While it is very unlikely Ross can guess the content of the DNS cookies (each real cookie is a 64-bit pseudo random string), Ross could tap Josh's phone and learn what the secret words (cookies) are and still manage to order a pizza to be delivered to Josh. Such on-path attacks, however, are relatively rare compared to off-path attacks, against which DNS cookies are quite effective. DNSSEC, on the other hand, avoids this issue entirely by providing end-to-end validation that the answer came from the source and that it hasn't been tampered with during transit.

DNS cookies are not a complete fix. They are more of a Band-Aid. But they make a pretty good Band-Aid. Because all they require is that you turn them on, they are simple to implement.

Summary: Cache Poisoning and DNSSEC

Despite advances in security, cache poisoning remains a valid threat against recursive resolvers, especially Internet-facing ones. It is attractive to threat actors because once successfully executed, cache poisoning is very difficult for DNS administrators or security operators to detect. Although the Internet community has added some fixes to address the original vulnerability, they are only temporary measures because the fundamental lack of security in DNS is still there. In the meantime, security professionals can focus on preventing cache poisoning by adding restrictions on the recursive resolvers and analyzing unsolicited DNS responses. Slowly, the Internet is moving toward DNSSEC, which offers the true fix for cache poisoning. In the meantime, DNS cookies are already available and are easier to deploy than DNSSEC, even though they don't offer complete protection from cache poisoning like DNSSEC does.

CHAPTER 8

Encrypted DNS

Before You Read This Chapter

Some of the content in this chapter could be out of date by the time you read this book. We have seen rapid development in the area of encrypted DNS in the recent past, and we are sure to see even more in the coming years. Here we provide a general overview of the concepts, allowing security professionals to make informed decisions and designs using these evolving technologies.

In previous chapters, we have outlined techniques for detecting DNS-based attacks. Those techniques depend on our ability to inspect DNS queries to look for evidence of malicious activity. Unfortunately, with the arrival of encrypted DNS, that crucial ability may be taken away from security professionals, leaving us unable to protect users from threats that exploit DNS. What's worse, chances are, users on your network are already employing encrypted DNS right now because it is being deployed by default on numerous browsers. While the intention behind encrypted DNS is to provide added privacy, some implementations inadvertently generate serious security implications.

Encrypted DNS: Protecting the Last Mile

Encrypted DNS is an industry attempt to address DNS abuse and privacy in a systemic way. It has been championed by browser providers and users alike. Encrypted DNS offers no protection against cache poisoning. It does, however, protect DNS communications to some degree. Here is a basic overview of what it is along with trade-offs to consider in its use and deployment.

Perhaps you are a very forward-thinking security professional and have already enabled DNSSEC validation on your recursive resolvers, turning them into validating resolvers. Congratulations! Even though a majority of Internet domains have not yet deployed DNSSEC, at least you have secured the server-to-server communication for DNS as much as possible.

However, the segment between the validating resolver and the client device (stub resolver) remains wide open to hacking. This unprotected segment is also known as the DNS "last-mile" problem. The gist of the problem is that even when clients visit only DNSSEC-signed domains (there are only a few of those to date), and your validating resolver is verifying every single DNS response, the final delivery of that answer to the client is still sent in the clear without any confidentiality or authentication of the response contents. As Figure 27 illustrates, an attacker sitting between the client and the validating resolver can snoop and perform man-in-the-middle attacks.

We can protect this last-mile communication with encrypted DNS. However, keep in mind that this solution encrypts only the communication between the client and its configured resolver. It does not provide end-to-end encryption for DNS resolution—that is, the communications between DNS servers are not encrypted;

encryption is only from DNS server to DNS client. As we'll see, this is only a partial fix that includes security trade-offs.

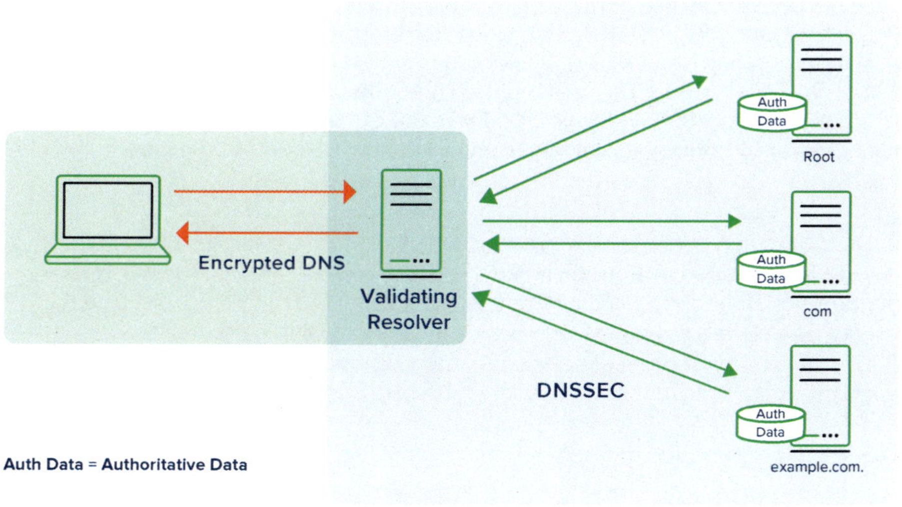

Figure 27: DNS last mile

Leading Encrypted DNS Standards

The idea behind encrypted DNS is straightforward — use the power of encryption so DNS messages are safe from prying eyes, including those of threat actors and companies hoping to monetize user data without express permission. Public key cryptography, such as transport layer security (TLS), is a technology well-suited for this purpose because we already use it for encrypting most of our communications. There's one major catch: TLS needs TCP.[65] Because most DNS communications run over UDP, this poses a challenge, and it is a major reason why DNS encryption has been slow in coming.

The DNS community recognized that privacy is a concern it could no longer avoid, so efforts started a few years ago to add encryption support. In recent years, several competing standards have emerged. We will describe two of the most dominant technologies, DNS over TLS (DoT) and DNS over HTTPS (DoH), and discuss how they impact the jobs of security professionals, along with the security implications of each.

In the context of discussing DoT and DoH, we will refer to the "regular" DNS as DNS over port 53 (Do53, which could potentially be pronounced as "dose").

65 Technically, TLS over UDP is possible with DTLS, as described by RFC 6347, but it isn't widely supported today.

DNS Over TLS (DoT)

The design principle for DNS over TLS (DoT)[66] is simple: Make no changes to the DNS message format, just move everything from UDP port 53 to TCP port 853 and add TLS encryption on top of it. Mostly due to the requirement of using a new TCP port, deploying DoT requires significant coordination and planning. In addition to having a DoT client and a DoT server, all of the network devices along the path must open TCP port 853. However, DoT has the advantage for administrators in that it is easier to control and block, making it better suited for on-premises private networks.

With DoT, administrators can configure clients to automatically switch to DoT when available (called opportunistic encryption). This means DoT clients can be configured to fall back to Do53, if there is not a suitable DoT server. Alternatively, the client can have a strict profile, where it must send DNS communications only over DoT; if no DoT servers are available, the client will not perform any DNS resolution.

DNS Over HTTPS (DoH)

The design goal for DNS over HTTPS (DoH)[67] from the start was to make DoH messages indistinguishable from other HTTPS messages. Using DoH is simple for most users because many web browsers and some operating systems already support DoH. Users can easily enable DoH on the client, point to any public DoH server on the Internet, and start using DoH because most networks permit port 443 outbound by default (it's the same port as HTTPS).

While it is easy to enable, routing DNS traffic in this way poses significant problems for security practitioners. The main challenge is that DoH messages cannot be distinguished from other HTTPS traffic. In other words, when using DoH, DNS-specific threat activity cannot be detected, controlled, or blocked. In fact, this is one of the reasons why red teaming tools, such as ColbaltStrike and Silver, make extensive use of DoH.[68]

DoH does not offer opportunistic encryption like DoT, but most implementations automatically fall back to Do53 when DoH resolution is unavailable. DoH can offer an additional layer of privacy through Oblivious DoH (ODoH).[69] ODoH addresses the privacy concern that when using encrypted DNS services, the DNS server can see both the IP address of the client and the DNS content. That places a

66 DNS over TLS (DoT) is specified in RFC 7858.

67 DNS over HTTPS (DoH) is specified in RFC 8484.

68 https://outflank.nl/blog/2018/10/25/building-resilient-c2-infrastructues-using-dns-over-https/

69 Of course, it is worth considering if the additional complexities introduced with ODoH are worth the effort or are even a problem worth solving. Unless clients are going to start running their own recursive DNS servers, they will have to trust someone at some point to provide secure, reliable resolution services.

lot of trust on the encrypted DNS server to do the right thing. ODoH sends DoH messages through a proxy, which then forwards them on to the DoH server. As Figure 28 shows, the proxy knows the IP address of the client but cannot see the message, and the DoH server sees the message but does not know who the client is.

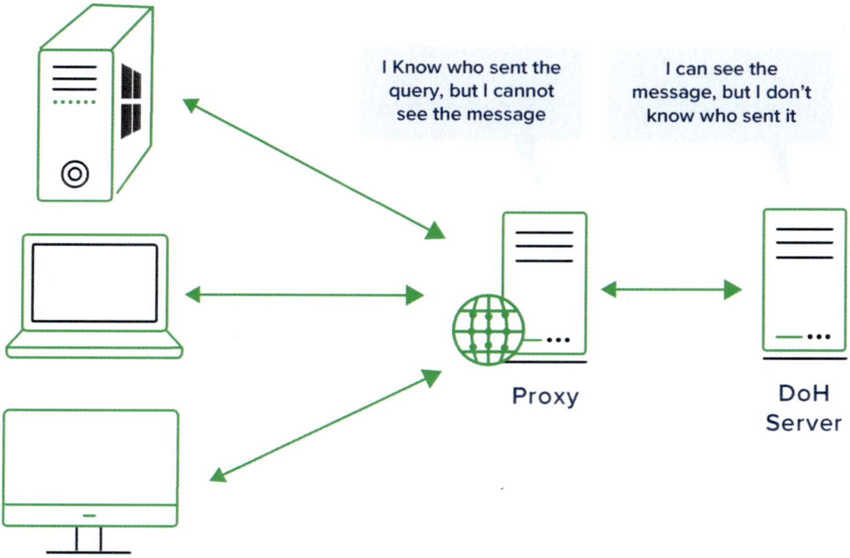

Figure 28: Oblivious DoH

Greater User Control Adds to Security Complexity

Encrypted DNS technologies grant users not only more privacy but also more control. Most encrypted DNS implementations allow users to override the DNS settings configured for the operating system, while some offer more granular controls, such as per-application behavior or even per-domain behavior. However, as we'll learn in a moment, depending on the implementation, greater user control may create a false sense of privacy, while also making it more difficult for enterprises to enforce security policies.

Let's see how encrypted DNS impacts Alice, a security-conscious user. She reads about encrypted DNS and decides that she is better off with it turned on, so she enables DoH on her Android mobile phone and uses the public DoH server provided by Google. This action means that now all DNS traffic travels via TCP port 443 to Google's public DoH server, 8.8.8.8. Then, Alice changes the app behavior for Firefox so that all DNS queries from Firefox are sent to CloudFlare's 1.1.1.1 DoH server. Furthermore, she finds a setting that allows her to change the lookup by domain, so she sets up a rule that all lookups for the domain `example.com.` will be resolved using the Quad9 DoH server, 9.9.9.9.

Alice feels confident that she has control over her DNS privacy, and her ISP cannot resell her DNS data for profit. But is this true? Let's look at this from the administrator's point of view.

First, the additional privacy Alice believes she's achieved is an illusion. Her DNS data, instead of being visible to her home ISP, is now going to companies including Google, CloudFlare, and Quad9. These companies are most likely doing a lot more analysis and reselling of her DNS data than her home ISP did. Alice's setup might work well for her at home, but when she goes to work and gets on the corporate network, she might not be able to resolve internal domain names, such as `ldap.company.local.`, among others. The complex setup on her mobile device will be very difficult to troubleshoot. Lastly, because all her DNS communications are encrypted, the security team at Alice's workplace cannot see what she is looking up or enforce any type of DNS security policy.

Encrypted DNS appeals to end users with the promise of privacy, but makes it difficult for the security team to enforce policies and for the infrastructure team to maintain and troubleshoot. Even worse, by encrypting the DNS traffic, we have provided additional cover for attackers who seek to misuse DNS, as we will see later in this chapter's case study.

Public Encrypted DNS Services and Privacy Trade-Offs

DoH is quickly gaining traction, mostly due to its ease of installation for end users. Starting in 2020, browsers such as Firefox and Chrome have DoH enabled by default. Mobile operating systems, such as iOS and Android, both support DoH and DoT natively. Even Microsoft has added DoH support for its operating systems Windows 10 and Server 2022. These default deployments mean many users are using DoH without realizing it, sending DNS queries to public DoH providers. We thought it would be interesting to security-minded readers to compare how these popular online DoH providers handle user privacy data. The following is a summary of what we found in our policy reviews:

- *Google:* Google Public DNS is such a common and popular service now that even non-technical users know to use its IP address, 8.8.8.8, for DNS resolution. In June of 2019, Google announced support for DoH. The privacy statement[70] notes that Google collects personal information, but does not use it for targeted ads, nor does Google correlate or associate personal information from Google Public DNS with a user's other Google services (e.g., Gmail), except for addressing security and abuse. Google keeps two types of logs, temporary and permanent. The temporary log contains both the source IP address and query and is deleted automatically within 48

70 https://developers.google.com/speed/public-dns/privacy

hours. The permanent log replaces the source IP address with city or region location but retains other information such as ECS (EDNS client subnet) and AS (autonomous system) numbers.

- *Cloudflare:* If you downloaded Firefox in North America, this is one of the default providers enabled in the browser. Cloudflare offers its public DoH service on the IP address 1.1.1.1. Cloudflare supplies several tiers of plans, from free for home users to paid plans for businesses and enterprises. According to its privacy policy,[71] Cloudflare does not collect personally identifiable information and collects only partial IP addresses.[72] The bulk of the data is stored for 24 hours. Cloudflare shares the scrubbed data with APNIC[73] for research purposes.
- *Cisco:* The networking company formerly known as OpenDNS is now part of Cisco and falls under its Umbrella line of services. Similar to Cloudflare, OpenDNS offers different tiers of plans, although its home version requires users to sign up for an account. Cisco's privacy policy does not explicitly state what details are tracked or stored. However, according to the Umbrella Privacy Data Sheet,[74] it does track personal data, sometimes down to the internal Active Directory information, made evident by this statement: "Because Cisco Umbrella processes, stores, and analyzes DNS, web and full traffic depending on package and deployment, and where applicable, processes and stores active directory information, it processes certain personal data of the users." Cisco claims the collected data is used to predict threats and enhance threat intelligence for its customers.
- *Quad9, or 9.9.9.9:* The non-profit public resolver, Quad9 Foundation is headquartered in Switzerland, which means it is subject to European privacy laws, such as the General Data Protection Regulation (GDPR). In addition to providing free encrypted DNS services, such as DoH and DoT, Quad9 also filters malicious domain names at no charge, when many other vendors offer it as a paid service. As for privacy data collection,[75] Quad9 increments counters based on source country, region, and AS number but not individual IP addresses. Quad9 does collect aggregated data for the DNS query and shares it with "a few carefully vetted security researchers" for threat intelligence research.

Our aim here is not to single out any provider. We merely want to demonstrate that not all encrypted DNS providers are created equal, and not all of them provide the same level of privacy protection for the end users. When it comes to encrypted DNS providers, we hope the takeaway is *know your privacy rights and read the fine print.*

---

71 https://www.cloudflare.com/application/privacypolicy/

72 Last octet of IPv4 addresses and last 80 bits of IPv6 addresses

73 Asia Pacific Network Information Center https://blog.apnic.net/2018/04/02/apnic-labs-enters-into-a-research-agreement-with-cloudflare/

74 https://trustportal.cisco.com/c/r/ctp/trust-portal.html?search_keyword=umbrella#/1552559092863133

75 https://www.quad9.net/privacy/policy/

A Double-Edged Sword (Don't Cut Yourself!)

Encrypted DNS gives users privacy from certain network administrators, but this can also be used by malicious actors to hide their evil actions. Without encrypted DNS, network administrators can see what users are doing (less privacy). In turn, that enhanced visibility helps protect users from malicious networks and destinations. In contrast, with encrypted DNS, network administrators cannot see the DNS queries (more privacy). However, that enhanced DNS privacy comes at a cost. It also prevents security teams from using DNS-focused security measures such as RPZ to safeguard network communications. The next case study highlights that threat actors are already taking advantage of the ways that encrypted DNS enables them to hide their presence.

Case Study: GodLUA

```
Classification: Malware
Observed: 2019, months after DoH RFC was approved
Tactics: Encrypts C2 communication in DNS over HTTPS, then
uses other methods to download malware payload
```

During 2019, just a few months after RFC 8484 for DNS over HTTPS was approved, NetLabs[76] observed the first malware that (ab)uses the new protocol. The malware has a few novel techniques, including a redundant C2 communication mechanism that involves using `pastebin.com.` and `github.com.` in its execution. For our purposes, we will focus on how the malware leverages DNS.

Early versions of GodLUA relied on unencrypted DNS lookups for the initial contact to find C2 servers then switched to other channels such as HTTPS to download malicious payloads. Later versions of GodLUA added the capability to download malicious payloads directly via DNS TXT records. Because the latest version of GodLUA is using DoH, now all of these messages are exchanged as encrypted HTTPS messages.

The authors or group behind GodLUA were confident that their tracks were covered by DoH, so much so that they did not bother to set up look-alike names or register DGA domains. Rather, GodLUA used a hard-coded domain name, `d.heheda.tk.`, and later added the `cloudappconfig.com.` domain. Figure 29 shows an example[77] of a DoH message using a TXT record to download a malicious payload.

76 https://blog.netlab.360.com/an-analysis-of-godlua-backdoor-en/

77 Source: https://blog.netlab.360.com/content/images/2019/06/dnstxt.PNG

```
GET /dns-query?name=t.cloudappconfig.com&type=TXT HTTP/1.1
Host: cloudflare-dns.com
Accept: application/dns-json

HTTP/1.1 200 OK
Date: Wed, 26 Jun 2019 10:22:25 GM
Content-Type: application/dns-json
Content-Length: 345
Connection: keep-alive
Access-Control-Allow-Origin: *
Cache-Control: max-age=214
Expect-CT: max-age=604800, report-uri=https://report-uri.cloudflare.com/cdn-cgi/beacon/expect-ct
Server: cloudflare
CF-RAY: 4ece75c228ebb120-HKG

{
  “Status”:0, “TC”: false, “RD”: true, “RA”: true, “AD”: false, “CD”: false,
  “Question”: [
    {“name”: “t.cloudappconfig.com”, “type”: 16}
  ],
  “Answer”: [
    {“name”: “t.cloudappconfig.com”, “type”: 16, “TTL”:214,
     “data”: “\”6TmRMwDw5R/sNSEhjCByEw0Vb44nZhEUyUpUR4LcijfIukjdAv+vqqMUYOFA
     oOpC7Ktyyr6nUOq09XnDpudVmbGoTeJD6hYrw72YmiOS9dX5m/spNmsw/eY/XDYYzx5/\””
    }
  ]
}
```

Figure 29: Sample GodLUA DoH message

Minimizing Risks From Encrypted DNS

Because of changes made by major browsers and OS vendors, we believe that eventually most enterprises will be forced to offer DoT or DoH services internally. Until the new infrastructure is in place, allowing corporate users to connect to external encrypted DNS servers is a big risk for the enterprise. As security professionals, we must plan to deploy and maintain our own encrypted DNS servers. This means you need to have the capability to operate and maintain your own DoT or DoH servers internally. If you don't, you *will* lose control of your DNS traffic: clients will query DNS servers you have no control over. You also can't inspect the messages (encrypted), thus you cannot enforce security policies that rely on DNS (such as RPZ). Until you have your own encrypted DNS servers, here is what you can do to mitigate risks associated with encrypted DNS:

- *Disable client OS and software DoT or DoH settings.* If possible, disallow or disable DoT or DoH settings on the client, so users cannot unwittingly turn it on. Some web browsers come with DoH enabled by default.
- *Restrict access to external encrypted DNS servers.* By blocking or restricting access to external DoT and DoH servers, you help ensure that no one from the corporate or enterprise network is bypassing security policies and talking to malicious DNS servers. For DoT, this is easy. Simply block TCP port 853. Unfortunately, it is much harder to block DoH, which we discuss next.

Blocking DoH

DoH has been designed to look identical to other HTTPS messages. The only way to tell if the HTTPS message is a request for a web page or a DNS query is to decrypt the traffic and perform a deep inspection of the actual HTTPS message. In addition, TCP port 443 is ubiquitous for Internet access; we cannot simply block port 443 and hope to selectively block DoH traffic. Instead, we have three possible tactics to control outbound DoH access: proxy, blocking public DoH servers (via IP address and name), and canary domains.

Proxy

Your organization may already have a traditional web proxy (or transparent proxy) set up. If so, you can configure clients to force them through the proxy, letting you inspect and block the encrypted DoH messages.

Block Public DoH Servers

The second tactic is to block HTTPS access to public DoH servers. Take the popular Google Public DNS server IP address, 8.8.8.8, for example. If users are trying to connect to this IP address via port 53, we know that traffic is Do53 (regular DNS). However, there is no web content hosted on that IP address. That means that if anyone is trying to connect to 8.8.8.8 over TCP port 443, we know that to be DoH traffic, and we should block it. Many security vendors offer a curated list[78] of publicly known DoH (and DoT) servers, so you can add them to the blocklist on your security devices. Such block lists are also available from the RPZ perspective as well.

In many cases, DoH uses Do53 for the initial name resolution (a process known as bootstrapping) to locate the actual DoH server and then establish encrypted DNS communications. Google Public DNS's URLs use the names `dns.google.com.` or `dns.google.`, for their DoH servers. So, if you block resolution of those domain names over Do53 (in conjunction with other techniques such as blocking public DoH servers by IP address), you effectively block the DoH connection from forming in the first place. You are then able to force clients to use Do53 infrastructure that is within your control.

Canary Domains

The third tactic, first proposed by Mozilla and adopted by other major browser vendors, is called canary domains.[79] The idea is simple. Set aside a special domain name (e.g., `use-application-dns.net.`) as the DNS "kill switch" for DoH. The web browser performs a DNS lookup for the canary domain. If the answer is NXDOMAIN, it informs the browser to stop using DoH and fall back to using Do53

78 This is a basic (not complete) list of public encrypted DNS resolvers: https://dnsprivacy.org/public_resolvers/.

79 https://support.mozilla.org/en-US/kb/canary-domain-use-application-dnsnet

(regular DNS over port 53). Figure 30 illustrates how canary domains work with web browsers.

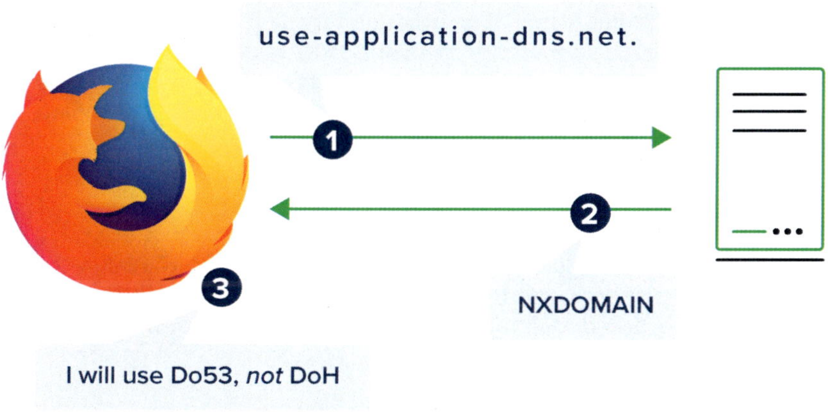

Figure 30: Canary domain

1. A web browser such as Firefox queries for the canary domain name `use-application-dns.net.`.
2. The DNS server has been configured to return NXDOMAIN to disable DoH.
3. The web browser uses system-defined DNS settings instead of DoH.

DNS Last-Mile Features Comparison

To help you visualize the security features offered by each last-mile technology discussed so far, Table 4 lists features they offer. Do53 is the traditional DNS with no security features. Some of you may question why DNSSEC, which provides authentication for server-to-server communication, is listed in the table even though it is not a last-mile technology. Because Do53, DoT, and DoH are all last-mile technologies, they cannot provide authentication of the DNS response data. That's what DNSSEC adds to the mix.

In other words, in order to get the full spectrum of security benefits including authentication, you must deploy DNSSEC, regardless of which last-mile solution(s) you choose.

Is it...	Do53	DNSSEC	DoT	DoH
Encrypted?	X	X	✓	✓
Easily Blocked?	✓	N/A	✓	X
Authenticated?	X	✓	X	X

Table 4: DNS last-mile features comparison

Encrypted DNS Considerations

It may seem like we are being overly critical of encrypted DNS as a technology by pointing out all of its shortcomings. That is not our intent. Our goal is to call attention to potential pitfalls so security professionals can avoid them. In mid 2022, encrypted DNS looks like it is definitely here to stay, with DoH gaining more market share. There are even rumors about auditors recommending or requiring the use of encrypted DNS. It makes sense. DNS is one of the few remaining technologies still running over plaintext on the modern network, where everything else has already been encrypted.

This section offers areas to consider when deploying encrypted DNS. We do our best to describe in general terms, knowing the specifics will likely be different by the time you read this section due to rapid development in this area.

Hop-to-Hop Behavior

As you design and deploy encrypted DNS, keep in mind that DNS is a per-hop technology. What that means is, how each DNS server talks to the next-hop DNS server depends on how the two servers are configured. You could have data encrypted from point A to B but plaintext from point B to C.

Performance

Both DoT and DoH add overhead with the use of TCP and TLS. Both technologies are too new for us to say definitively what their performance impact might be. Nevertheless, initial research[80] suggests that DoH performs about the same as Do53, while DoT can outperform both Do53 and DoH. These results may come as a surprise, but when you consider that there are TCP techniques such as pipelining multiple queries in a single TLS session, the superior performance of DoT is understandable. A new standard, DoQ (DNS over QUIC),[81] sends DNS messages over the QUIC[82] protocol for even more performance gain. In short, performance for encrypted DNS is likely not a major concern.

Deployment and Discovery

As of summer 2022, if you were to deploy your own DoT and DoH servers, you would need to perform manual key management and distribution tasks. That is, DoT and DoH clients must either already have the server's public key or certificate hard coded in or they must be manually imported. This lack of automation makes it cumbersome for mass deployment. There are two proposed standards in the

80 https://pschmitt.net/docs/www2020-dns-dot-doh.pdf

81 https://datatracker.ietf.org/doc/rfc9250/

82 QUIC is specified in RFC 9000.

works to address those challenges, Discovery of Designated Resolvers (DDR)[83] and Discovery of Network-designated Resolvers (DNR):[84]

- DDR allows clients to locate the "official" encrypted DNS servers for the network with Do53 by resolving specific names.
- DNR relies on a new DHCP option to tell network clients where to find the encrypted DNS services for the network segment.

As of this writing, we cannot determine which standard will become more widely used, so we recommend would-be implementers keep their eyes on both DDR and DNR.

Can You Trust the Resolver?

By using encrypted DNS, we have privacy between the DNS client and server (the DNS last mile), but there is no mechanism to verify that the DNS server is trustworthy. For example, users could enable DoH and start encrypting all DNS traffic to a remote DoH server owned by a threat actor, who obtained a valid TLS certificate through reputable vendors, such as VeriSign or Let's Encrypt.

Just as with HTTPS, we can check the validity of the TLS certificate, but the *trustworthiness* of the target is much more difficult to determine. Experienced security professionals understand the difference between privacy and trustworthiness; one does not necessarily equate to the other. This distinction needs to be communicated clearly when considering encrypted DNS. Just because it is encrypted does not mean you can trust it. The question you should ask yourself is, *do I trust public Internet encrypted DNS servers more than my own?*

---

83 https://datatracker.ietf.org/doc/draft-ietf-add-ddr/

84 https://datatracker.ietf.org/doc/draft-ietf-add-dnr/

Summary: Privacy and Encrypted DNS

Encrypted DNS was going to happen sooner or later. While many users cheer the arrival of encrypted DNS and the privacy benefits it brings, security professionals know privacy is a blade that can cut both ways. If not implemented properly, encrypted DNS will impair the security team's visibility and control in the environment. As a security professional, you should collaborate with the DNS infrastructure team and find out what encrypted DNS solutions fit your business use cases. Aim to provide your own encrypted DNS services, rather than relying solely on public encrypted DNS services.

Public encrypted DNS servers might be a reasonable choice for home users, but in the enterprise, they are rarely a good fit. More clients—from desktop OSs, to mobile platforms, to web browsers—are supporting DoT and DoH natively. Over time, more are coming with encrypted DNS features enabled by default. We can no longer ignore encrypted DNS in the enterprise,[85] and we must have a strategy to manage it.

85 For an example showing rapid adoption of encrypted DNS in various entities, see the United States Federal Zero Trust Strategy, released in early 2022. It recommends the use of encrypted DNS: https://www.white-house.gov/wp-content/uploads/2022/01/M-22-09.pdf

CHAPTER 9

DNS Attacks Against Clients

When Client Devices Are in the Crosshairs

Security teams typically have an impressive lineup of tools and processes to secure client devices, from user permissions to agent software, and everything in between. However, when it comes to threats intended to compromise clients or endpoints, DNS vulnerabilities are often overlooked. In this chapter, we will look at how DNS clients have evolved over the years, what threats are targeting each client component, and tactics that security teams can employ to better protect clients from DNS-based threats.

Before and After DNS Resolution

The DNS attacks we've discussed thus far in this book occur during DNS resolution, as client devices contact DNS servers to retrieve answers to DNS queries. There are, however, a few threat tactics that target clients before they contact DNS servers and others that unfold after resolution. Figure 31 illustrates the steps a typical client with a web browser goes through before contacting a DNS server. We will see shortly how threat actors can exploit these steps to carry out attacks.

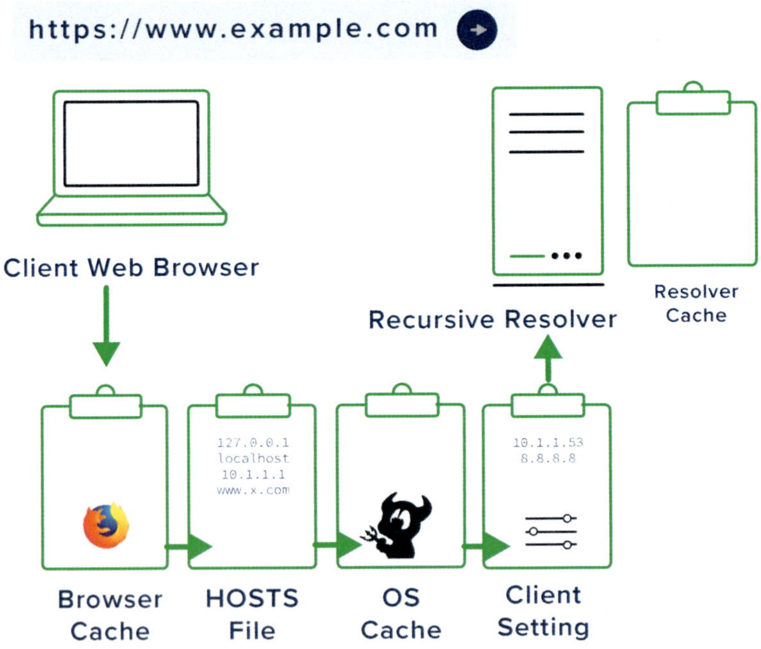

Figure 31: Before contacting DNS server

Our example walks through the client using a web browser to access the web resource `https://www.example.com.`:

1. First, the client application, a web browser here, checks to see if the web page or the domain name `www.example.com.` is already in the browser cache.
2. If it's not there, the client checks the local HOSTS file to see if it contains any hard-coded domain names.
3. If the domain name is not found in the HOSTS file (or if the HOSTS file does not exist), the client checks the operating system for any potential OS-level DNS caching.
4. Finally, if no entries are found in the OS-level DNS cache (or the OS does not cache, such as Linux), a DNS message is sent to the configured DNS server(s) according to the client DNS settings, which can be manually set or provided by DHCP.

Keep in mind that the order presented here is typical but not definitive. Some devices or applications may not follow the same order. The specifics will depend upon the application and OS configuration.

After the client has contacted the DNS server and received an answer, almost the same process happens in reverse (Figure 32). Cache entries are shown as red lines on the pages in the figure. The recursive resolver that retrieved the answer stores a copy in its own cache before passing it on to the client. The client stores the answer in its various caches. The answer may be stored in the client OS cache and the browser cache. It is unlikely the client will make changes to the HOSTS file. As we will see in the case study in a minute, bad actors count on the assumption that HOSTS files are rarely changed and are processed before DNS resolution.

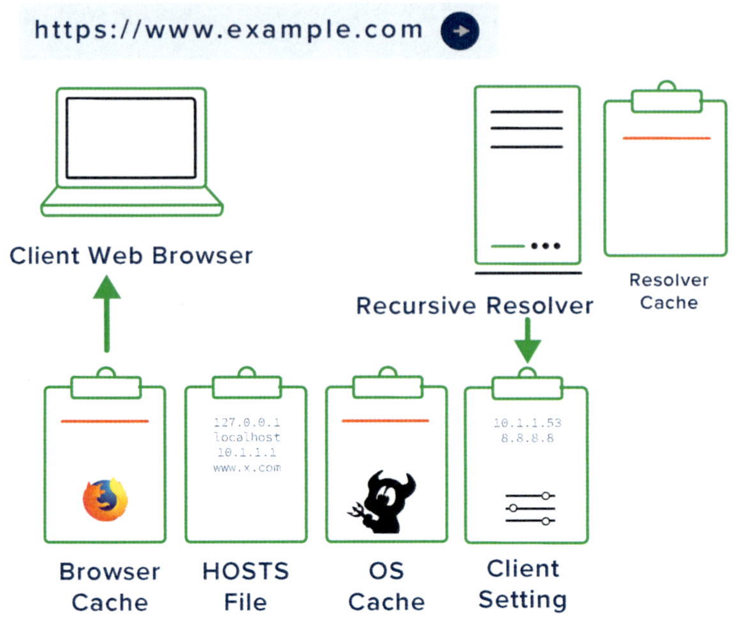

Figure 32: After contacting DNS server

Three Types of DNS-Based Client Attacks

Because it is refreshed so frequently, the browser cache is generally not a target for threat actors. However, other forms of cache stored on client devices are very much at risk. In this section, we will learn more about how attackers may target each of these components on client devices: HOSTS file, OS Cache, and client DNS settings. Note that in the explanations that follow, the attackers exploit these files that are used before DNS resolution occurs.

Important: Securing these client components is crucial. You can deploy the most robust DNS servers on the market, protect those servers with DNSSEC, and use encrypted DNS, but if attackers manage to compromise any one of these areas on a client's device, that client is vulnerable to attack because it will look at them first for answers before reaching out to a DNS sever.

HOSTS File

The HOSTS file predates DNS. In fact, DNS was created because HOSTS files could not scale along with the exponential growth of the Internet. In the early days of the Internet (or as it was known then, the ARPANET), if computers needed to look up an IP address for a given name, it checked a local file named "HOSTS." The content of the file looks very similar to a DNS name entry (though without the trailing dot), usually something like this:

```
127.0.0.1      localhost
192.168.1.2    x.example.com
```

Even after DNS was in wide use, many systems kept the HOSTS file as a type of local override to DNS entries, and it is often still used in that capacity today. For *nix-based systems, this file is usually located at /etc/hosts; for Windows-based systems, it is usually %WinDir%\System32\Drivers\Etc\hosts. The HOSTS file grants administrators some flexibility when configuring clients, but it also provides a new attack vector for threat actors, as the following case study shows.

Case Study: Win32.QHOSTS

```
Classification: Trojan
Observed: 2009
Tactics: Modifies hosts file to redirect popular domain names
to download malware
```

Win32.QHost or Win32/QHOST is a trojan that targets Windows devices in a certain geographic region. It alters the HOSTS file by adding entries for popular domains to attacker-controlled IP addresses. Furthermore, it adds entries to prevent victims from accessing security vendor domains. Figure 33 shows an example of how the trojan prevents victims from ever accessing security vendor domains by inserting entries like these into the HOSTS file, sending all queries for security vendors back to the device's own loopback IP address (127.0.0.1), effectively disabling any security updates. Figure 34 is an example of malicious entries that redirect victims from popular domains to an attacker-controlled IP address (192.157.49.9).

```
127.0.0.1 avp.com
127.0.0.1 ca.com
127.0.0.1 customer.symantec.com
127.0.0.1 dispatch.mcafee.com
127.0.0.1 download.mcafee.com
127.0.0.1 f-secure.com
127.0.0.1 kaspersky.com
127.0.0.1 liveupdate.symantec.com
127.0.0.1 liveupdate.
symantecliveupdate.com
127.0.0.1 localhost
127.0.0.1 mast.mcafee.com
127.0.0.1 mcafee.com
127.0.0.1 my-etrust.com
127.0.0.1 nai.com
127.0.0.1 networkassociates.com
127.0.0.1 rads.mcafee.com
127.0.0.1 secure.nai.com
127.0.0.1 securityresponse.
symantec.com
127.0.0.1 sophos.com
127.0.0.1 symantec.com
127.0.0.1 trendmicro.com
127.0.0.1 update.symantec.com
127.0.0.1 updates.symantec.com
127.0.0.1 us.mcafee.com
127.0.0.1 viruslist.com
127.0.0.1 viruslist.com
127.0.0.1 www.avp.com
127.0.0.1 www.ca.com
127.0.0.1 www.f-secure.com
127.0.0.1 www.kaspersky.com
127.0.0.1 www.mcafee.com
127.0.0.1 www.my-etrust.com
127.0.0.1 www.nai.com
127.0.0.1 www.
networkassociates.com
127.0.0.1 www.sophos.com
127.0.0.1 www.symantec.com
127.0.0.1 www.trendmicro.com
127.0.0.1 www.viruslist.com
```

Figure 33: Win32/QHost HOSTS file sample 1

```
192.157.49.9 my.mail.ru
192.157.49.9 m.my.mail.ru
192.157.49.9 vk.com
192.157.49.9 ok.ru
192.157.49.9 m.vk.com
192.157.49.9 odnoklassniki.ru
192.157.49.9 vk.com
192.157.49.9 www.odnoklassniki.ru
192.157.49.9 m.odnoklassniki.ru
192.157.49.9 ok.ru
192.157.49.9 m.ok.ru
192.157.49.9 www.odnoklassniki.ru
```

Figure 34: Win32/QHost HOSTS file sample 2

OS Cache

RFC protocol standards for DNS do not specify the client implementation for cache, and OS designers and implementers are free to choose whatever DNS behavior they think is best. In some operating systems, such as Microsoft Windows, DNS entries are cached locally,[86] while others, such as Linux, do not. Many administrators prefer to have a local cache for the perceived performance gain, but as we will see below, this gives the attackers yet another path to bypass security controls.

Case Study: Dridex

```
Classification: Trojan, malware
Aka: Bugat, Cridex
Observed: Since 2009
Tactic: Targets banks using undocumented Windows API to
perform local DNS cache poisoning
```

Originally observed in 2009, the Dridex trojan targets banks and financial institutions. Since 2016, it has evolved to target some cryptocurrency wallets.[87] Dridex sends its victims attachments, usually Microsoft Word or Excel files. When opened, macros download components for Dridex and attempt to redirect victims, sending them to malicious destinations that impersonate legitimate banking institutions. One technique includes abusing an undocumented Microsoft API DnsAddRecordSet_A to inject entries into the local DNS cache.

86 When DNS entries are cached by DNS servers, they are generally stored for a time specified by a value known as time-to-live (TTL). However, some clients do not honor the TTL duration, so the information may be retained for a longer or shorter time than the namespace's administrator intended. Experienced Windows users will be familiar with the command ipconfig /flushdns, which clears the local DNS cache.

87 https://news.softpedia.com/news/dridex-banking-trojan-will-soon-target-crypto-currency-wallets-508041.shtml

This type of cache poisoning does not affect multiple clients like DNS server cache poisoning in Chapter 7, but it can still be devastating. To detect this type of infection, we would have to monitor the client's DNS cache, probably with a locally installed agent. Some security professionals are proponents of disabling client-side caching altogether, and this type of attack makes a strong case for it.

Client DNS Settings

If a client has checked all its local resources (web browser cache, HOSTS file, and OS cache), and still cannot find the domain name it wants to resolve, then it is time to reach out to a DNS server. But how does the client find a DNS server? Or more specifically, how does it find a recursive resolver? Many people mistakenly refer to a client's recursive resolvers as primary and secondary name servers. In DNS, the words *primary* and *secondary* refer to specific roles of the authoritative servers. These client implementation terms are not part of DNS specifications; however, the proper terms are *preferred* and *alternate* as Figure 35 shows.

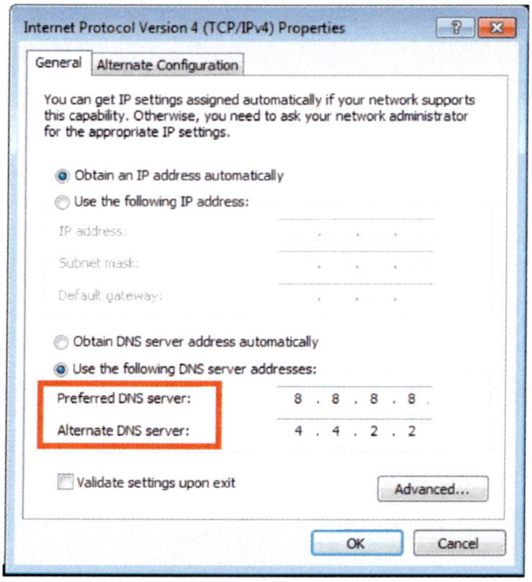

Figure 35: Windows preferred and alternate DNS settings

The user or the administrator can manually configure this setting. The settings can also be established by the OS as part of the DHCP process.[88] This next case study shows how threat actors exploit these settings to carry out attacks against clients.

88 DHCP Option 6 contains DNS server IP addresses.

Case Study: DNSChanger

```
Classification: Trojan
Observed: 2011
Tactic: Modifies computer's DNS preferred and alternate name
servers setting to use rogue recursive resolvers, redirects
users to advertising sites
```

DNSChanger is an old but well-known trojan. It infects victims by changing the DNS settings to use rogue recursive resolvers that malicious actors control. This trojan effectively changes the preferred and alternate DNS server addresses shown in Figure 35, and it replaces the IP addresses with the ones controlled by the bad actors. By redirecting a client's DNS recursive queries, the attacker essentially has full control over the client. For instance, the client could query for the domain name `www.microsoft.com.`, and the attacker's recursive resolver would return an IP address they control. At its peak, DNSChanger was estimated to have infected over four million computers. Luckily, the threat actors behind DNSChanger were merely displaying fraudulent advertising pages for profit; the attack could have been so much worse.

The lesson of this case study is that enterprises should always control the flow of DNS traffic. A common best practice is to block any TCP/UDP port 53 outbound traffic from anything except DNS servers your organization controls. If you are allowing clients on the enterprise network to connect to any external DNS recursive resolvers, with or without encrypted DNS, you could fall victim to another DNSChanger-style attack.

The Evolving Client Landscape

We have taken a brief look at some of the possible attacks against DNS clients. However, not all clients are equally vulnerable. We have categorized DNS clients into four types, with each requiring its own security control measures.

Classic Clients

Classic clients are well-known operating systems on servers, desktops, and laptops. They include Microsoft Windows, Apple's MacOS, and various flavors of Linux and Unix variants. Configurations and behaviors for these clients are well documented and widely understood, thus they are easier to manage from a security perspective. The behavior of these clients is also generally predictable. For example, when the preferred name server becomes unreachable, the Windows 8 stub resolver will generally contact an alternate name server after one to two seconds.[89] Classic clients are our baseline. Everyone is familiar with them, and we have well-established tools and tactics to secure these clients. So,

89 https://learn.microsoft.com/en-us/troubleshoot/windows-server/networking/dns-client-resolution-timeouts

let's move on to the emerging client categories, because they require additional security considerations.

Mobile Devices

Mobile device behavior is newer, less well understood, and less consistent than classic clients, making them harder to securely manage. This category includes devices such as mobile phones, tablets, and smart watches. Some of these devices cache DNS results more aggressively to preserve power. Often these behaviors are not documented, or the configuration is unavailable to administrators. What's more, the behavior could change frequently from vendor to vendor, sometimes even from version to version. Another security concern about mobile devices is that they often switch to different network profiles, with each profile exhibiting different DNS behaviors. For example, when a mobile phone switches to LTE connectivity, it may store entries in its DNS cache more aggressively than when it is on a Wi-Fi network. To make matters worse, many of these mobile devices do not provide a way to manage or change DNS settings for their cellular connections. This means less control and visibility for the security team, as these mobile clients switch to and from different network connections.

IoT Devices

The Internet of Things (IoT)[90] has entered not only our homes but also the enterprise. From doorbells, televisions, and security cameras to printers, HVAC systems, and commercial lighting, these devices are even less well understood and less configurable than mobile devices. The vast majority of IoT devices do not expose their DNS configuration or behavior. Some even use completely different DNS protocols known as multicast DNS (mDNS) and DNS-SD,[91] a kind of peer-to-peer DNS. Many IoT devices are manufactured by companies inexperienced in computer security and are designed to reside only on home networks. In addition, most IoT devices are full of security vulnerabilities that are difficult or impossible to patch, and their DNS behavior is also drastically different than other clients and cannot be changed.

Web Browsers

Web browsers have come a long way since the late 1990s and have evolved into miniaturized OS platforms. Browsers typically interact with the OS through system calls, such as `gethostbyname()`, which does not return TTL values. Thus, browsers have no choice but to arbitrarily choose a TTL value, and most choose somewhere between 30 and 60 seconds. The latest development in browsers is the addition of DNS over HTTPS (DoH) capability, which allows web browsers to ignore DNS settings at the OS level and make their own DNS queries over HTTPS,

---

90 Or as the authors refer to them: Insecure Distributed Internet of Things (IDIoT).

91 DNS-SD is used by some vendors to perform local or cloud-based service discovery.

potentially bypassing enterprise security policies. A browser performing its own resolution would then be able to see the TTL values returned from the DoH resolver. We discussed encrypted DNS in detail in Chapter 8.

Options for Protecting Clients

Fortunately, when it comes to defense, we have more options for the clients. You are probably aware of the swarm of endpoint protection software that is already looking for suspicious files and changes to system files. To prevent attacks like Win32.QHost, you can add the client HOSTS file to that list of system files.

As we mentioned in Chapter 7, DNS caching is a double-edged sword. It can reduce lookup times, but when the cache content is tampered with, it creates a whole new class of threats that are very difficult to detect. Client OS-level caching is especially vulnerable. To bolster protection, some operating systems, such as most Linux distributions, do not cache DNS at the OS level. Some people argue that this adds strain and potential lag time to the local DNS infrastructure. Others maintain that eliminating client-side cache (or limiting it to very short timeframes, such as a few minutes) while continuing to rely on DNS-server-level cache could be a reasonable compromise between performance and security.
As for the client's DNS server settings, most clients receive their settings via DHCP. Describing ways to protect the DHCP server and its transport is beyond the scope of this book. However, you can monitor the configuration changes on the client to ensure that the client is communicating only with DNS servers permitted by policy. Such measures help prevent a client from using a DNS server controlled by a threat actor.

Summary: DNS Attacks Against Clients

When it comes to securing DNS, most protection efforts are focused on the DNS resolution process. DNS security at the client device level is often overlooked. Security practitioners must be mindful of threats that target such client components as the HOSTS file, OS cache, and DNS settings. In addition, among categories of client devices, mobile, IoT, and Internet browsers each may require additional security measures. In this age where network clients are ever-more mobile and remote, we need to account for users not only within the enterprise premises but also users connecting via VPN, working from home, or connecting from a public Wi-Fi network at a coffee shop.

Some of you might wonder: Why haven't we talked about hijacking? Isn't that a common threat against clients? For that, we need a whole chapter, which is coming up next.

CHAPTER 10

Domain Hijacking

When the Domain Name Itself Is the Target

Attackers are all about exploiting the weakest link. In this chapter, we focus on the domain name itself. Domain hijacking exploits weaknesses that often fall off the radar. Examples include a domain name you registered but forgot to renew, an application server you decommissioned last year, or a domain name registration company you work with. Read on to learn how domain hijacking works and tactics to prevent it.

What Is Domain Hijacking?

Before we begin, let's agree on terminology. *Domain hijacking* has varying definitions depending on who you ask. For the sake of clarity, we break down the various forms of domain hijacking and give them more descriptive names. Whatever form domain hijacking takes, the overall concept remains the same: The threat actor gains administrative control of a domain name that belongs to someone else.

What Domain Hijacking Is Not

As we discuss the particulars of domain hijacking, it's helpful to know what it is not. Some mistakenly consider RPZ rewrite or substitution as a type of domain hijacking. In this attack, users send a query for domain "X," but DNS servers with an RPZ policy for "X" rewrite or substitute it with answers for domain "Y." Many captive portals work this way. Unauthenticated clients send a query for a common website, such as `www.google.com.`, but are redirected to the `payment.portal.local.` site.

This technique is better known as DNS redirection, DNS doctoring, DNS substitution, or DNS rewriting. We covered this tactic in the RPZ section in Chapter 3. This is not domain hijacking because the legitimate domain name still exists. The response data from the legitimate domain has simply been substituted for a different one in the response returned to the client. In true domain hijacking, nefarious actors gain control of the legitimate domain name itself. In addition, not all DNS redirects are performed with malicious intent. In fact, many PDNS (protective DNS) service providers offer redirects as a way to prevent users from connecting to dangerous domain names. Because this technique mainly affects recursive resolver behaviors and not authoritative domain names, we do not consider it part of the "DNS hijacking" or "domain hijacking" family, even though some people may refer to it as such.

Domain Squatting

Another illegal activity sometimes confused with domain hijacking is domain squatting. In this form of abuse, bad actors register domain names using trademarked words and phrases from legitimate businesses and organizations with the intent to charge victims exorbitant fees to acquire them. Remedies for this abuse generally involve legal actions associated with trademark infringement or ICANN's UDRP[92] process, but those topics are outside the scope of this book. That said, security-aware organizations would do well to perform trademark searches on a routine basis to ensure that no one is using branded terms in domain names without authorization.[93]

How Domain Hijacking Attacks Unfold

Malicious actors have several options for perpetrating domain hijacking attacks. Some of these take advantage of the structural processes involved in registering domain names while others exploit the DNS asset management complexities inherent in cloud deployments. In the sections that follow, we describe each of these attack pathways—domain name registration infrastructure compromise, expired and dormant domains, and record abandonment—in detail and provide examples where we can. In some cases, there are variations within the category, in which case we also call out those differences.

Domain Name Registration Infrastructure Compromise

Chances are, seasoned security professionals such as yourself have already implemented a consistent policy for domain registration, renewal, and approval. You probably know exactly what domain names your organization currently uses and can document who is responsible for each one. Good for you! The next question is, how safe is your registrar? Some of you may even be wondering what a registrar is. Before delving further into the category we call "domain name infrastructure compromise," we first need to review the basics of how domain registration works.

Domain Registration Basics

First and foremost, domain registrars are organizations that officially register your domains. They are external to your organization, with their own sets of security processes and policies. Domain registrars are pertinent to any DNS security discussion because while you may have the most stringent defensive measures in place, your registrar may not. Furthermore, your registrar could, in fact, be your weakest security link.

---

92 UDRP stands for Uniform Domain-Name Dispute-Resolution Policy; you can read more about it here: https://www.icann.org/resources/pages/help/dndr/udrp-en

93 There are also businesses that offer this type of monitoring as a service, such as MarkMonitor, DomainTools, and CSC, among others.

Figure 36 below shows the different entities for address and domain name registrations. The critical pieces for our discussion are the lowest level registrants (that's us) and the one just above it on the right, registrars (the organizations registrants pay to register a domain name).

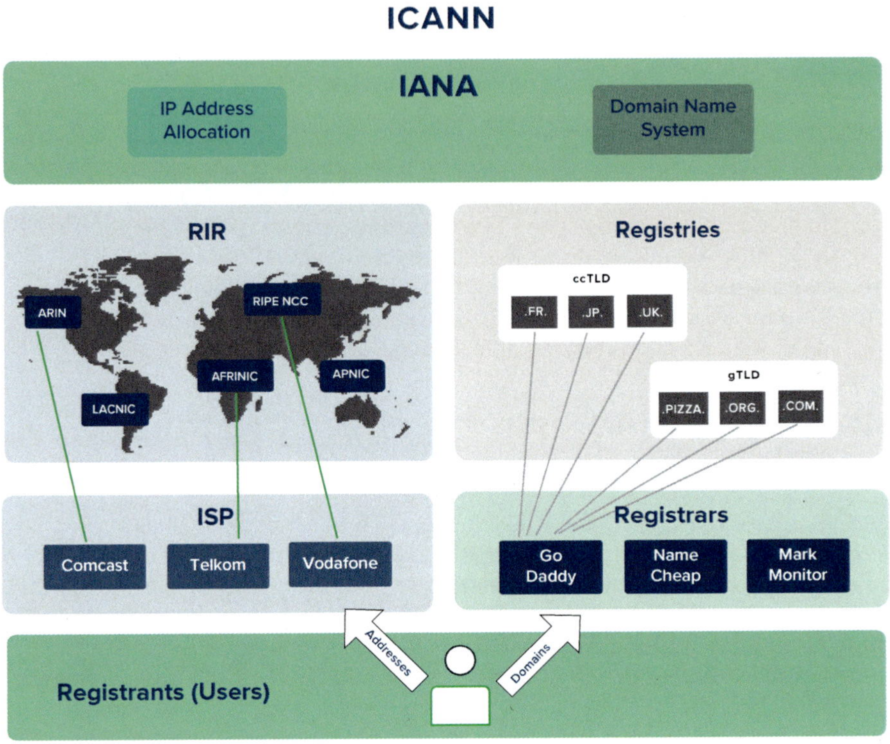

Figure 36: Hierarchy of domain operational authorities

Name registration generally goes like this:

1. User pays registrar, such as GoDaddy, for a domain name, usually for a recurring annual fee.
2. Registrar asks the user to provide NS (such as `ns1.example.com.`) and glue record information (such as 10.1.1.1).
3. User sets up authoritative DNS servers at the IP addresses provided in the glue records.
4. The registrar interacts with the registry that runs the TLD DNS servers (such as Verisign for `com.`) to publish the NS and glue records on those DNS servers.
5. Voilà! The world can resolve the domain name now!

For the domain example com., Figure 37 shows an example of its NS and glue records: ns1.example.com. and ns2.example.com. are the NS record targets. The glue records are the IP addresses for these names, 10.1.1.1 and 172.16.2.2, respectively.

Figure 37: Sample domain registration

Once the registration is complete, whenever Internet users want to resolve anything that ends with example.com., recursive resolvers of the world will be able to follow referrals from root to com. and from com. to example.com. ("contact ns1.example.com. at 10.1.1.1 or ns2.example.com. at 172.16.2.2") and ask one of the returned name servers to receive an authoritative answer.

How Attackers Strike

Remember that a domain registrar has its own security processes and policies independent from yours. Many registrars use custom software for their management interface, which could have vulnerabilities that attackers can exploit.

Attackers can also use social engineering tactics to alter NS or glue records. A common tactic is to call the registrar's helpdesk and provide a sob story: "I am new to this job. The last DNS administrator quit, and we need to change these resource records today! Please help me!" If a threat actor can change the NS record or the glue record, she basically gains control of your entire domain by setting up her own authoritative name server and returning any answer she wants.

Furthermore, the login to manage the domain is often a shared account. Many users don't treat this account's login credentials with the same level of respect they may have for their day-to-day network logins. The result often is weak or shared passwords, which threat actors can exploit. Once a threat actor gains access to your account, she can change domain registrations any way she pleases.

What makes this form of attack even worse: It is difficult to detect quickly. DNS traffic to your authoritative name servers will not disappear all of a sudden. It will likely taper off over time as cached entries expire from recursive resolvers, and traffic will migrate to the attacker-controlled name server. You won't know that your domain has been hijacked in this manner until it's too late.

Hosted DNS Data Compromise

Many organizations today do not even bother with setting up their own DNS servers. Instead, they outsource this activity through DNS hosting, a service many DNS registrars and some other third parties offer. In such instances, the DNS registrars are not merely brokers between domain names and users but also holders of all DNS data. Today, it only takes a few minutes for any user to register a domain name and add a few records, without setting up an authoritative DNS server. All of them will magically appear on the Internet. Thanks to DNS hosting services, the tasks involved in creating and managing DNS data are easier than ever before.

As readers of any book with the word "security" in the title, we all know where this discussion is going. Before the age of hosted DNS services, attackers perpetrating hijacking had to go to the trouble and expense of setting up their own DNS servers. They would then need to breach the registrar to change NS and glue records to point to their malicious DNS servers. Thanks to the convenience of hosted DNS services, attackers just need to gain access to the management account and change any resource records they want. No additional servers necessary! Want to apply for a new TLS certificate? Change the CAA record. Want to disrupt DNSSEC? Remove or change important DNSKEY records. Want to redirect emails? Change the MX records (as this next case study shows).

Case Study: Fox-IT

```
Classification: Domain hijacking
Observed: 2017
Tactic: Attacker changed hosted authoritative DNS records,
performed man-in-the-middle attacks
```

Fox-IT is a cyber security company based in the Netherlands. As a security expert, Fox-IT certainly keeps a close watch on its own infrastructure and activities. Threat actors chose to compromise its domain registrar, which had less strict security controls, and altered the DNS records to perform a man-in-the-middle attack. Fox-IT was able to detect and remedy the compromise in roughly 10 hours and limit its overall damage. Few organizations would be so lucky. Here's how the attack unfolded.

After threat actors compromised the domain registrar, they had full access to all DNS resource records. They changed MX records to control the flow of emails. Next, they used the ill-gotten email access to obtain a legitimate TLS certificate. The threat actors then set up a web server with the legitimate TLS certificate and again changed DNS records to send users to an unauthorized web server. When users visited the web page, they did not suspect anything was wrong. The DNS records seemed legitimate, and the web server had a legitimate TLS certificate. Figure 38 illustrates the Fox-IT domain hijacking incident.

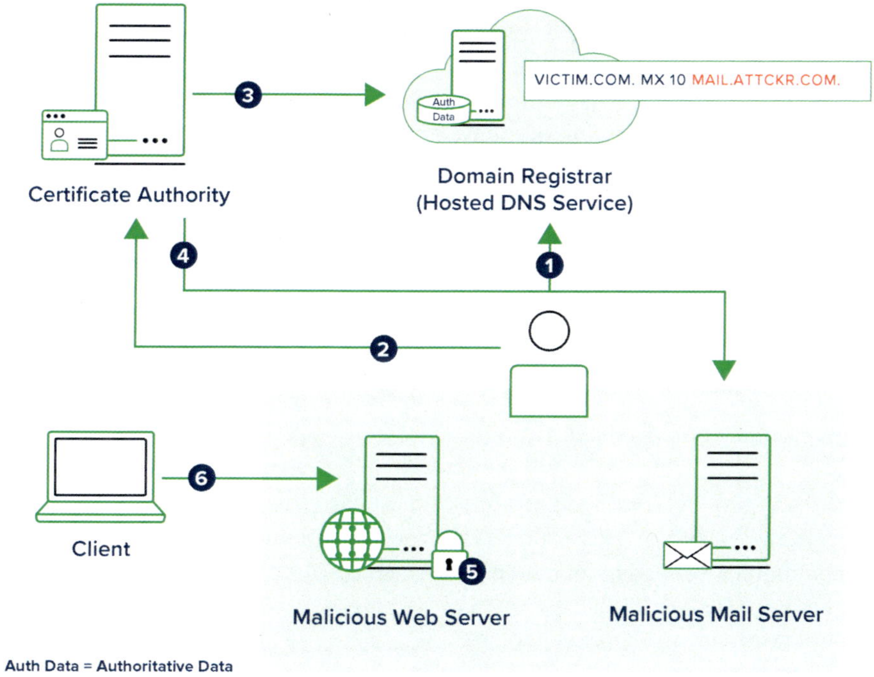

Figure 38: Fox-IT domain hijacking

1. Attacker changes MX record at the hosted DNS provider.
2. Attacker requests a new TLS certificate.
3. Certificate authority (CA) looks up MX record in order to send verification email.
4. Attacker answers verification email and receives TLS certificate.
5. Attacker installs legitimate TLS certificate on malicious web server and alters DNS to publish web server address.
6. Client connects to impostor web server.

The lesson we can all learn from Fox-IT is that your domain registrar's security needs to be part of *your* defense strategy. In the case of Fox-IT, it even went as far as having deployed authoritative DNSSEC. But the attackers were able to bypass all layers of security, including DNSSEC, by simply breaching the weakest link, the registrar itself.

Fox-IT detailed more information about this incident in a blog entry,[94] and one of the most important takeaways was: Choose a domain registrar that offers multi-factor authentication (MFA).

Rise in Recent Years

On January 22, 2019, the U.S. Department of Homeland Security (DHS) released an emergency directive[95] warning of the type of attack in the Fox-IT example, giving it yet another name: "DNS Infrastructure Tampering." This announcement came as the result of hackers tampering with many U.S. federal government domains.[96] Around the same time, we also saw similar threats, such as DNSpionage[97] and Sea Turtle,[98] which performed mass domain hijacking at a global scale.

The DHS emergency directive warned readers of the threats we have outlined above: Attackers can breach DNS accounts, alter DNS records, obtain certificates, and redirect email as a result of the breach. The document recommended four actions that organizations can take and that are more important than ever:

1. *Audit DNS records:* This is a good ongoing practice that assures you that important DNS records were not altered without authorization. There are commercial services that can monitor your DNS records and alert you when they change.
2. *Change DNS account passwords:* DNS registration or hosting account credentials are usually not under as much scrutiny as regular network accounts. You should subject them to at least the same standards and require users to practice good cyber hygiene.
3. *Add multi-factor authentication to DNS accounts:* Not all registrars offer MFA. For those that do, enable it; for those that don't, consider migrating your domains to another registrar that does.

---

94 https://blog.fox-it.com/2017/12/14/lessons-learned-from-a-man-in-the-middle-attack/

95 DHS Emergency Directive on DNS Infrastructure Tampering https://cyber.dhs.gov/ed/19-01/

96 https://www.mandiant.com/resources/global-dns-hijacking-campaign-dns-record-manipulation-at-scale/

97 https://blogs.cisco.com/tag/dnspionage

98 https://arstechnica.com/information-technology/2019/04/state-sponsored-domain-hijacking-op-targets-40-organizations-in-13-countries/

4. *Monitor certificate transparency logs:* While certificate transparency (CT) is not strictly DNS related, it provides useful information regarding which certificates were issued for the domain. This information can help identify whether the domain has been compromised.

Choosing a Secure Registrar

When it comes to choosing a domain registrar, most people probably don't give it much thought. Most registrants just run a quick Internet search for the name and then opt for whichever registration company offers the name at the cheapest price. That's the extent of their registrar selection process.

As we have shown in this chapter, your organization's domain registrar is part of your security strategy, like it or not. Would you sacrifice security to save $5.99 a year? Of course not. There are several things to look for in choosing a secure domain registrar. At the top of the list is requiring MFA. Below are additional criteria to help you choose a better, more secure domain registrar.

- *Auto-renewal option:* Gone are the days when domain registrars waited for your domain to expire, then contacted you to pay to keep your domain name. This practice was once a common way for less-than-reputable registrars to sell you back the same domain name, often at a higher price. Look for registrars that offer auto-renew options, to save you from having to remember to renew the domain every year and risk having malicious actors take over the domain.
- *Privacy features:* WHOIS[99] is one of the oldest protocols on the Internet. Traditionally, when you register a domain name, you need to provide two contacts, one administrative and one technical. WHOIS publishes this information for everyone to see. Nearly 40 years later, this level of transparency is often undesired. Look for registrars that offer privacy services to hide your personal information, displaying generic information instead of specific, private information. Some registrars charge additional fees for the privacy service; some include it as part of the standard package.
- *DNSSEC support:* You have learned about DNSSEC in Chapter 7. Currently, not all registrars support DNSSEC-signing (that is, the authoritative side of DNSSEC). If you are serious about DNS security, look for a registrar that supports DNSSEC, and plan your DNSSEC deployment.
- *Additional security features:* Some registrars provide a registration- or record-locking feature to prevent accidental (or illicit) changes. Some offer monitoring services to detect changes in critical resource records. Others may alert you when similar-looking domains are registered (such as look-alike domains described in Chapter 3). Ask around! You might be surprised by what other features are available.

---

99 The WHOIS protocol was originally published in RFC 812 in March 1982, making it officially older than DNS, which was not proposed until 1983.

On-Premises DNS Infrastructure

While we're on the topic of infrastructure compromise, some of you might wonder why we have yet to discuss threats related to on-premises DNS servers anywhere in this book. On-premises servers are clearly part of the infrastructure, but alas, they are not a focus of this book. The reason why is that protecting on-premises DNS servers is a broad subject well beyond our scope with techniques and considerations that vary depending on the architecture, design, configuration, and version of DNS servers and their configuration. We hope to cover this subject in a future book.

Before we move on to the other two categories, we need to say a quick word about how cloud infrastructure adds to hijacking challenges because it has bearing on the remaining two pathways in this discussion.

DNS Security in the Cloud

As seasoned network and security professionals, we sometimes joke about the cloud: "It's just storing data on computers that someone else owns." There is some truth to that statement when it comes to DNS security. DNS-based hijacking that takes place in the cloud focuses not on the specific DNS implementations or weaknesses in the protocol but on the DNS management process (or lack thereof).

Cloud computing has many benefits, including the ability to dynamically provision and deprovision resources to match demand. However, if DNS entries are not integrated into the provisioning and deprovisioning process, we could end up in a situation where the DNS records are pointing to a resource, but that resource is no longer active.

In a traditional on-premises enterprise environment, losing track of DNS records in this manner does not pose too much of a threat. For example, users might occasionally complain that they can look up the domain name but cannot access the website. However, in a cloud environment, especially a shared environment, where names and IP addresses are recycled and reused, attackers can take advantage of these abandoned DNS names.

Now that we have that out of the way, let's get to the discussion of expired and dormant domains along with record abandonment.

Expired and Dormant Domains

Imagine you rent a building for your candy store to fulfill your childhood dream. The building has everything you could possibly want—a convenient location, beautiful architecture, and ample space for your business to grow.

Then imagine, one day, you arrive in the morning only to find that the lock has been changed! Not only that, but an impostor is in your shop, running your business, talking to your customers, and nobody else can tell the difference! This must enrage you. How did it happen?

You look back in your piles and piles of mail and find out that you forgot to renew the lease for the building. Someone else swooped in and snatched up the lease, and now this terrible person is conducting business under your name.

Of course, this doesn't happen often in physical retail spaces. But it happens more often than you might imagine in cyberspace with domain names. If you forgot to renew your domain name lease (usually every year), there are plenty of threat actors who are paying close attention, waiting to snatch up your domain name. Technically, they did not break any rules or laws. They waited until your domain name registration expired before registering it for themselves. But this is morally dubious at best and, very often, outright malicious. Some of these threat actors just want to make some quick cash and will sell you back your domain. Others will enjoy the residual reputation of your domain name and cause you much more pain.

You might be thinking: What kind of idiot forgets to renew the building lease or domain name? Unfortunately, it is much more common than you might think. Consider this for a minute. Which individuals or teams in your organization are responsible for registering domain names and renewing them? The answer heavily depends on your organization. It might range from legal to marketing to the IT department, or all of the above. In many cases, the larger the organization, the murkier this responsibility becomes, especially if individual business units start registering domain names on their own (which should be prohibited for many reasons, including this one).

A simple and pertinent exercise is to collect all the domain name registrations that currently belong to your organization. For some, it will be an eye-opening experience to see just how many different domain names there are in your organization, which registrars are associated with the various names, who in your organization was responsible for registering them, and how many of those individuals have left the organization.

The number one reason domain name registrations lapse is from an absence of formalized internal registration processes. As security practitioners, we should establish clear rules for the organization when it comes to domain registrations. Start by answering the following questions:

- *Which individuals or teams should approve domain registration?* If possible, establish clear criteria for domain name registration, such as business need and approval from a manager.

- *Which domain names should you register with which registrars?* Registrars act as brokers between the top-level domain registry and users.[100] There are many Internet domain name registrars. Some offer competitive prices and others provide security features, such as MFA. Some registrars cater to the enterprise market and offer additional services, such as monitoring the misuse of an owner's trademarks.
- *Which email address(es) should be associated with domain registration?* Domain registrars require at least two email addresses, an administrative and a technical contact. These email addresses should be email distribution lists (such as hostmaster@example.com) and not an individual employee's email address (such as ross@example.com).
- *How will domain registration renewal be handled?* Most reputable domain registrars will send email reminders to renew a domain that is about to expire. You should have a process in place to handle such reminders. That process will enable you to decide whether to keep paying for the domain name, change the registrar, or let the registration lapse (and possibly be registered by someone else).

The following case study shows what happens when organizations fail to consistently use stringent domain name registration procedures.

Case Study: GoDaddy and Spammy Bear

```
Classification: Dormant Domain Hijacking
Aka: Bomb spammers
Observed: 2018
Tactic: Attacker breached registrar, commandeered dormant
domains to send out spam emails making bomb threats and
demanding a ransom
```

In December 2018, many users received what appeared to be legitimate emails from well-known corporations, including Facebook, Expedia, Mozilla, MasterCard, Hilton International, McDonald's, Yelp, Warner Bros., and DigiCert. The emails made threats to blow up schools, hospitals, public places, and businesses across North America, unless a ransom of $20,000 was paid. This caused a public panic, but ultimately no bombs were uncovered, and no physical harm occurred.

What's puzzling is, these emails looked like they came from legitimate senders. Further investigation revealed that these were all domains registered through GoDaddy, and many of these were dormant domains owned by corporations or organizations. No one was paying attention to these domains. Threat actors (likely

---

100 The distinction in roles between registrars and the registry for a given TLD is outside of the scope of our discussion here. In short, for any given TLD, there is only one registry, but many registrars can work with that registry to help sell names within the TLD.

the Spammy Bear group) found a weakness in the GoDaddy validation process and commandeered these forgotten domains.

Hundreds, possibly thousands, of domains were compromised. Take Mozilla's domain, `virtualfirefox.com.`, for example. From 2013 to 2017, this domain name resolved to its intended IP address of 74.80.210.168 (managed by a service provider based in California). Then on December 13, 2018, Farsight Security's data[101] shows this record was changed to 194.58.58.70, an IP address with a history of malicious content (based in the Russian Federation).

Any medium-sized enterprise likely owns dozens of domains. Many of these domains were registered at one point for a project, maybe placed on auto-renewal for the next few years, and subsequently forgotten. Well, we may have forgotten about them, but the threat actors have not. They are actively scanning and hunting for these forgotten or orphaned domains.

Abandoned DNS Records

As with many other DNS-related security threats, this attack has various confusing names, including, but not limited to, the following:

- Subdomain takeover
- Subdomain hijacking
- Use-after-free attack
- Dangling DNS pointer

We prefer to call this type of attack "record abandonment" because that's the root cause. Consider a typical cloud-based setup. You provision a new virtual instance. An IP address is dynamically assigned to it and sometimes even a DNS record is automatically created. However, that automatic record may not be in a namespace you control or be one that customers can easily remember. So, you create an alias that refers to that auto-created name.

The use of this record type, known as a CNAME, is common in hosted public clouds and content delivery networks. As an example, you might create a CNAME resource record for, say, `www.example.com.`, and point that record to something like `server-instance-use89c43f.region.cloudprovider.net.`, which would be the real A or AAAA record pointing to the IP address where the web page can be accessed. Figure 39 shows this typical DNS configuration.

101 https://krebsonsecurity.com/2019/01/bomb-threat-sextortion-spammers-abused-weakness-at-godaddy-com/

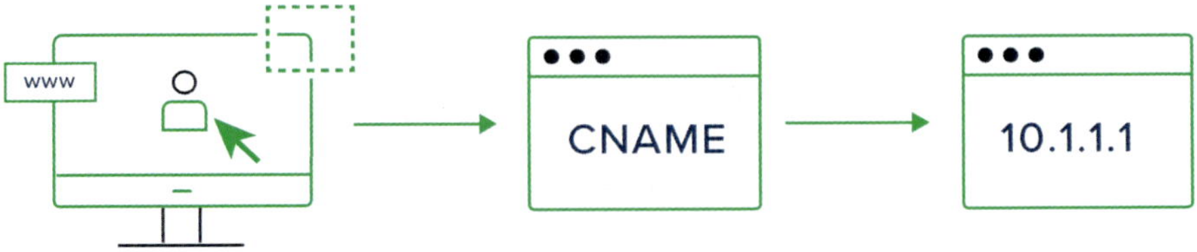

Figure 39: Typical DNS CNAME configuration

Now, let's say you are done with the project, and the instance or virtual machine is powered off. Mission accomplished. But you left the DNS entry in place. No big deal, right? Users' web requests to the resource will simply timeout because the web server is offline, right? That is no longer true. There are bots on the Internet today that scan for these abandoned DNS records[102] and then attempt to take over the IP address. This takeover is easier than it sounds, because the nefarious actors can just go to the cloud provider and ask for an instance. Cloud providers use a pool of IP addresses, and attackers will cycle through them until they receive the IP address they want—the one your DNS record is still pointing to. Figure 40 illustrates this behavior, with the name www abandoned, and the IP address 10.1.1.1 now controlled by the attacker.

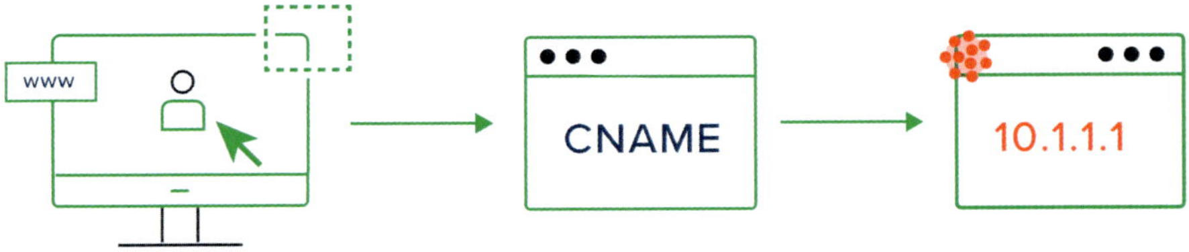

Figure 40: Hijacking IP address

In a variation of this attack, the threat actor focuses on subverting an intermediary CNAME record, often assigned by the cloud provider. An attacker can do this by exploiting the provisioning process of the cloud provider or performing a registrar compromise as described earlier in this chapter. In the end, the attacker controls the end target IP address (see Figure 41).

102 Researchers were able to receive the desired IP address in under 27 minutes, and estimated attackers can do so much faster. https://blog.apnic.net/2019/01/09/be-careful-where-you-point-to-the-dangers-of-stale-dns-records/

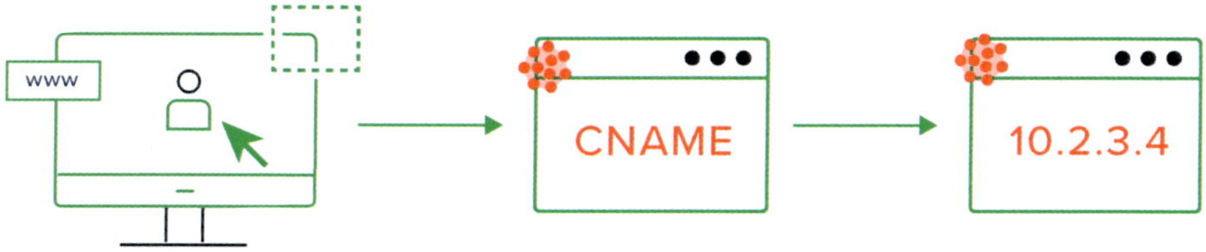

Figure 41: Hijacking CNAME

Our next case study shows you the repercussions of this form of attack.

Case Study: PowerDNS

```
Classification: Record abandonment
Observed: 2018
Tactic: Attacker obtained released IP address from provider.
Victim's abandoned DNS record supplied attacker with
residual reputation.
```

In September 2018, the team at PowerDNS foundation was briefly baffled by a warning from Google, notifying them that "hacked content" was found on the PowerDNS website. After some searching, they found the offending content on the `web-rtc1.powerdns.com.` site. They were scratching their heads because the server had been retired, and the virtual machine was disabled on their virtual private server (VPS) provider. The PowerDNS team then discovered that while the virtual machine was turned off, the VPS had reassigned its IP address to another instance that an attacker controlled. The attacker operated a web server complete with a matching certificate and was delivering malicious content under the legitimate domain name.

The PowerDNS team was able to fix the issue by removing `web-rtc1.powerdns.com.`, the stale DNS record. What's notable is how long it took to complete the fix. Because the name was already cached by many recursive resolvers around the world, the attacker enjoyed the residual reputation at least a few hours longer, if not a few days. So, while authoritative DNS servers can specify the TTL field for each DNS record, this value is sometimes ignored or overridden by an ISP or other providers, in order to preserve their own resources. Even if the PowerDNS team had configured the TTL to 300 seconds (5 minutes), in practice, this domain name is still cached in some recursive resolvers long past 5 minutes—often hours, sometimes days.

The takeaway: Keep your DNS data clean and up to date. And if you ever need to change a resource record, the change will take longer than you think to propagate.

What Can We Do About Record Abandonment?

The real issue here is accuracy—that is, your DNS data needs to accurately reflect your intended architecture. If a server or service has been provisioned, DNS data needs to send users to the correct IP addresses. By the same token, when a server or service has been removed, DNS data needs to stop resolving that name and sending users to an IP address that is not running the service. We can tackle this problem with three major approaches: *address management; automation;* and *auditing*.

IP Address Management

We are not talking about spreadsheets here. A proper IP address management (IPAM) system can help you organize not only what addresses are currently being used but also which were used in the past and which are planned for the future. Your provisioning team can use IPAM as part of the workflow to provision and deprovision, ensuring no IP addresses are left behind.

Automation

Modern provisioning and deprovisioning processes are often automated. In a cloud-based environment, this usually means making an API call to the provider to create or destroy a virtual machine. Make DNS a part of that process, so that when a new instance is created, the correct DNS entries are created, and when the instance is removed, corresponding DNS entries are also removed. If you have a strong IPAM solution in place already, this should be fairly easy to do.

Auditing

Experienced security practitioners know process and automation alone are not enough. We also need to conduct audits. Tools such as tko-subs[103] and Sublist3r[104] can help us actively hunt for orphaned or abandoned records, as can some monitoring services. If your DNS hosting provider or vendor can perform record auditing, that's even better.

---

103 https://github.com/anshumanbh/tko-subs

104 https://github.com/aboul3la/Sublist3r

Domain Hijacking Guide

	Cause	Description	Effect
1	Infrastructure compromise	Attacker gains control of registrar NS and glue records	Attacker can tell the world what my authoritative DNS server is
2	Infrastructure compromise	Attacker obtains control of hosted DNS data or access to the authoritative DNS servers	Attacker can change any of my DNS records
3	Expired domains	Attacker gains possession of domain registration during lapse	I forgot to renew my domain and now an attacker can register it instead
4	Record abandonment	Attacker gets control of the CNAME record for an application or system	Attacker enjoys residual reputation of my domain name
5	Record abandonment	Attacker gains control of IP address for an application or system	Attacker enjoys residual reputation of the IP address associated with my domain name

Table 5: Comparison of domain hijacking attacks

Summary: Domain Hijacking

Domain hijacking involves the use and abuse of domain names without permission. These attacks fall into three main categories: infrastructure compromise, expired and dormant domains, and record abandonment. Defensive techniques against various types of domain hijacking involve having clearly defined processes for domain name management and the handling of DNS records, requiring domain name registrars to follow defined security practices, and relying on IP address management in provisioning and security hygiene, along with periodic audits of the IPAM and DNS data.

CHAPTER 11

DNS and Zero Trust Architecture

In recent years, there has been increasing interest in the concept of the Zero Trust Architecture (ZTA), Zero Trust Network Architecture, or Zero Trust Network Access (ZTNA). In some places, it is also referred to as perimeterless security. As enterprises shift toward cloud infrastructure and as more team members work remotely, more enterprises are planning out a ZTA that meets their needs. We do not claim to be experts in this area; there are other books and resources dedicated to this topic. What we want to point out in this section is how you can leverage what you have learned from this book as part of your Zero Trust strategy.

Basics of Zero Trust

Zero Trust adds many entries to the already formidable IT dictionary. We are not going to use cool words and phrases like control plane, variable trust, or micro-segmentation here. Rather, we will focus on just three of the basics of a Zero Trust design:

1. Know where your assets are
2. Protect the data rather than just access to systems/networks
3. Use dynamic, session-based policy over static, address-based policy

In the following sections, we will show you how DNS (and other accompanying technologies) can help security practitioners prepare for implementing these areas of Zero Trust.

DDI Is the Foundation of Zero Trust

While there are many approaches to designing and building your new Zero Trust fortress, everyone seems to agree on what the foundation is. It is referred to by many different names, such as "defining your attack surface," "discovery of assets," or "inventory," but the ideas are the same. Simply put, before you can design a Zero Trust Network, *you need to know what assets you have and where they are.*

DNS these days is often mentioned as part of DDI (DNS, DHCP, and IPAM), and there are some very good reasons for this grouping. Let's take a closer look at each component.

- *DNS:* As we have seen throughout this book, DNS logs can provide not only what systems clients are trying to communicate with but also information about the clients themselves. For example, Firefox browsers will query for certain names when they first launch, iOS devices will query for certain sets of names when they boot up, and ransomware-infected hosts will query for certain names to establish C2 communication.

- *DHCP:* When DHCP is leveraged, lease history can provide the trail of a device's physical movements over time. DHCP can provide additional fingerprinting information and can act as a control mechanism to deny undesirable clients access to the network based on those fingerprints.
- *IPAM:* When IPAM is leveraged, it provides application and business contextual information, which in turn helps with risk assessment and event prioritization. An IPAM system can also be used to store security metadata, such as device patch level, owner contact, version number, and physical location.

In short, a solid DDI solution can act as a live asset inventory that is continuously updating. If your organization does not currently have an integrated DDI solution, we would strongly recommend that you explore your options in this space to provide you with the greatest visibility and control possible within your network.

DNS Can Detect and Mitigate Data Exfiltration

One of the core concepts of Zero Trust is to pay attention to defending data rather than access to systems. This means greater attention must be paid to data theft or data exfiltration, an area traditionally neglected by the perimeter model.

Figure 42 should look familiar to those of us who spend time working with network architecture and security. Most of us have spent hours in meetings and design sessions worrying about North-South traffic (inbound) and East-West traffic (lateral) movement. However, one direction that is discussed less frequently is the South-North (outbound) movement—that is, traffic leaving the premises and heading toward the Internet. Figure 42 illustrates these movements.

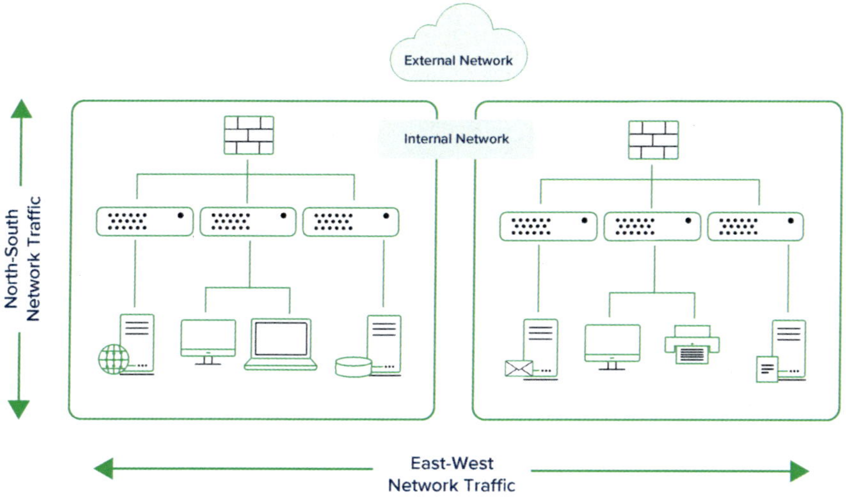

Figure 42: Data access model

Reports from Accenture, Cisco, SANS, and the NSA[105] all agree that over 90 percent of malware uses DNS to gain command and control, exfiltrate data, or redirect traffic. As we have seen in Chapters 5 and 6, tunneling and data exfiltration over DNS are commonly used by threat actors to evade detection and provide a reliable communication channel. This means *we should be tightly monitoring both in-bound and out-bound DNS traffic*.

We have covered the techniques on how to detect DNS tunneling and exfiltration attempts in Chapters 5 and 6. We have also covered the "silver lining" in Chapter 4, that simply observing the domain name can often help us identify which variant of the malware is active on the network. A good DDI solution should be able to take this information and update the IPAM database, for example, "Host 10.1.2.3. just queried for `ransom.example.net.`, a known domain name associated with ransomware XYZ. Update IPAM to reflect its status to 'potentially infected with XYZ'."

The next natural step is to limit the access of the potentially infected host. As we have seen in Chapter 3, we can achieve this with RPZ, a DNS-based technology. One obstacle for the security professionals in this area is the emergence of encrypted DNS technologies, described in Chapter 8. Prior to the emergence of DoH and DoT, it was pretty simple to limit access outbound on TCP/UDP 53 to only approved DNS servers. Unfortunately, those days are behind us, and especially in the case of DoH, it has become quite challenging to maintain control of all name resolution within a network.

We have a potential solution for that, described in the next section.

DNS-Enabled Dynamic Policy, an Introduction to D-NAP

Another of the core principles of a Zero Trust network is session-based dynamic policy. The idea is to have the policy dynamically update as the security posture of network hosts changes.

Enter the idea of DNS and Network Assured Policy (D-NAP).[106] The short version of describing D-NAP is integrating a firewall's policy with the managed DNS servers in a network such that any traffic to a destination that has not been resolved

105 For direct references, please see footnotes 2, 3, and 98

106 The concept of requiring DNS resolution before granting access isn't new, but we are proposing a different name for it here. A team from SIDN Labs posted a blog in September 2021 introducing the concept of what it called DNS Resolution Required (DRR). (See https://www.sidnlabs.nl/en/news-and-blogs/dns-resolution-required-can-help-make-the-internet-safer.) Unfortunately, with the emergence of DDR and DNR, this proposed name seems like it could be easily confused with the other technologies for discovering encrypted DNS resolvers. In addition, DRR doesn't seem to have caught on as a term yet, and other potential terms may emerge for this in the marketplace.

by the DNS server is blocked. D-NAP is a collaborative effort between the DNS server and the network/security infrastructure (hence the *and* in the name). In implementation, obviously there would need to be provisions for allowing certain traffic without DNS resolution, but we won't discuss that in detail here.

Figure 43 below illustrates the high-level flow of D-NAP.

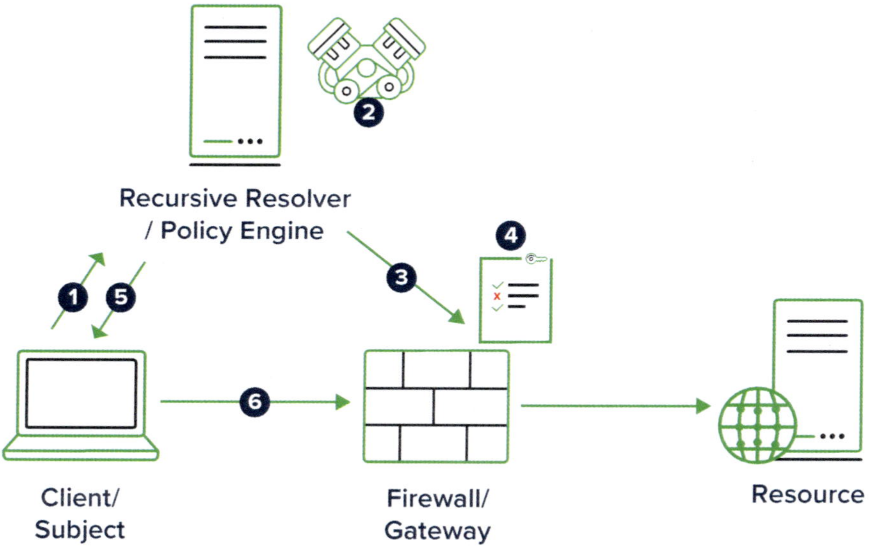

Figure 43: D-NAP illustrated

1. Client or subject sends a DNS query to the managed DNS recursive resolver, which is configured to enforce security policy (e.g., via RPZ).
2. The DNS server performs policy analysis of the query (i.e., if destined for a banned domain, execute configured policy action).
3. Assuming the response is allowed by policy and the name is resolved, the DNS server adds the response to its local cache and then communicates the resolved address to the firewall or gateway.
4. The firewall or gateway device, upon receiving the information from the DNS server, dynamically updates its access policy to allow the client/ subject temporary access to the resource.
5. The DNS recursive resolver sends the response to the client.
6. The client initiates its connection to the resource through the gateway or firewall, which is now allowed.

This may look like a lot of steps, but really it is not very different from what already happens today. The only additional steps are steps 3 and 4, where the information is shared with the gateway or firewall and where the access policy is dynamically updated. Although we listed steps 3, 4, and 5 as sequential steps, in implementation, they can be performed in parallel to minimize any potential added latency caused by introducing D-NAP to a network. We feel that D-NAP has the potential to be an essential part of security policy enforcement in networks built around a Zero Trust model.

So, what is the advantage of using a D-NAP approach? The main advantage is it provides the security professional with a way to ensure that all traffic leaving the network has either been explicitly approved or was vetted by the managed DNS server(s) on a per-session basis. This method, of course, assumes that the managed DNS server(s) have implemented threat intelligence technologies, such as RPZ.

In this session-based policy scenario, if a user uses an unsanctioned DNS server (a rogue DoH server, for example) to resolve the name `xyz.example.com.`, when the user attempts to establish the network connection, the connection would be denied by the gateway or firewall because all connections are blocked by default (according to the least privilege principle, also part of the Zero Trust Network philosophy).

We are not suggesting that encrypted DNS has no value in a managed environment. However, we do believe that it should be employed properly. Doing so means that clients should communicate only with encrypted DNS servers loaded with the appropriate policy (developed in conjunction with the security team) and monitored for any ongoing attacks or exfiltration attempts. This approach applies to regular DNS without encryption (aka Do53) as well.

Summary: DNS and Zero Trust Architecture

Being able to implement this D-NAP functionality is no small feat, considering that latency needs to be accounted for in each step along the way. Some of the potential increase in latency can be minimized by parallel processing and/or local caching. However, we believe these are reasonable trade-offs in the pursuit of a true Zero Trust environment.

CHAPTER 12

Conclusion

If you have read this book in sequence, then you should have a strong grounding in DNS and its security implications. As we stated at the outset, this book is not an exhaustive study of DNS. Our intention has been to give you the broad outlines of various ways that today's threat actors are able to exploit DNS's ubiquity and inherent vulnerabilities to carry out an ever-expanding range of cyber attacks. We've detailed some of the most prominent of these, along with steps you can take to defend against them. We have also attempted to point out security and privacy considerations involved in implementing various defensive measures.

We hope you've gained a deeper understanding of the hidden potential of DNS as a security control from this book. By now, it should be clear to you that by monitoring DNS communications you can detect malicious activities within the perimeter. This information provides rich data about endpoint clients and their behavioral patterns, as well as actionable items for remediation, such as blocking malicious domain names or taking infected clients offline.

If you are reading this book out of sequence, then the eight best practices that follow and the "Book at a Glance" bonus chapter can help you get oriented quickly.

Eight Ways to Fight DNS Insecurity Today

According to Zero Trust Architecture principles, nothing can be trusted on the network. In other words, we need to filter incoming traffic and pay attention to outgoing traffic, especially to new, unvetted domains. With these new "trust no one" and "guilty until proven innocent" approaches in mind, we have listed eight practical approaches for security practitioners to take in order to combat DNS insecurity.

1. Integrate DNS Into Security Operations

DNS should be tightly integrated into each step of the network and security life cycle. When provisioning and deprovisioning, making DNS part of the process will combat attacks, such as record abandonment described in Chapter 10. In day-to-day use, filtering by domain names can stop many (although not all) phishing, C2, tunneling, and exfiltration attempts. DNS itself can provide valuable information during threat hunting and incident response. Knowing what domain name was involved often can help security operations teams identify the active threat on the network.

2. Control, Log, and Monitor

The argument for logging and monitoring is easy to understand. The National Institute of Standards and Technology (NIST) provided seven tenets of Zero Trust in special publication 800-207, released in August 2020.[107] The seventh and final tenet states: *The enterprise collects as much information as possible about the current state of assets, network infrastructure, and communications and uses it to improve its security posture*. By observing the DNS queries made by network hosts, we can build a behavior profile and also get a fairly accurate sense of their security posture. Controlling the flow of DNS queries, through methods such as limiting which DNS servers clients can use to resolve names, is also important. Control of DNS queries makes collecting and monitoring query information easier, with or without encrypted DNS; if clients are getting answers from a rogue recursive DNS server, it spells trouble.

3. Use RPZ With Reputation Feeds

In Chapters 3 and 4, which explore look-alike domains and DGAs, we described how it is not feasible for security operators to manually keep up with the ever-growing lists of malicious domain names. RPZ enables recursive DNS servers to filter malicious domain names. Of course, such filtering depends on receiving quality threat intelligence feeds. Implementing RPZ is more than a passive protection; it can also protect users from such threats as look-alike and DGA domains. RPZ can be an essential tool to help defend against threat actors, allowing the security professional to proactively look for ongoing potential C2 communications or known malware active on the network and take remediation steps.

4. Perform Deep Query Inspection

In Chapters 5 and 6, DNS Tunneling and Data Exfiltration, we describe how blocklists have their limits, most notably when it comes to stopping zero-day threats or very targeted phishing or APT-style attacks. These limits are why the ability to perform deep DNS query inspection is necessary to help the security team analyze DNS query patterns and determine if an attack is in progress. We outline several key criteria that can be evaluated to help identify malicious DNS patterns. Armed with this knowledge, readers can write custom tools, purchase a commercial PDNS solution or service, or use a mix of both.

107 NIST Special Publication 800-207, Zero Trust Architecture, August 2020, available at https://doi.org/10.6028/NIST.SP.800-207

5. Audit DNS Data

As security professionals, we understand the importance of designing good policies and enforcing them. Auditing DNS data should be a part of the enforcement cycle. On recursive DNS servers, ensure that recursion is not open to potential threat actors. On authoritative servers, audit DNS records to avoid record abandonment and help prevent domain hijacking.

6. Deploy DNSSEC

While DNSSEC does not provide confidentiality, it does furnish authentication and integrity checking. Deploying DNSSEC on a recursive resolver can be as simple as pointing to one of the public DNS service providers already performing DNSSEC validation. Deploying DNSSEC on the authoritative side is more involved and requires careful coordination with domain registrars and parent zone operators, but it is a necessary step toward a global name system that can be fully authenticated. In the meantime, the already-deployed DNS cookies technology, while not a complete solution, offers reasonable protection against many attack scenarios.

7. Manage Encrypted DNS

Encrypted DNS is inevitable and to a large degree is already here. While many users cheer the arrival of easy-to-deploy encrypted DNS, this double-edged sword needs to be wielded carefully. Security professionals must understand the limitations of various encrypted DNS technologies and find the balance between privacy and control. For some, this balance means working with your DNS team to add encrypted DNS capabilities to your internal recursive DNS servers. For others, this means choosing one of the publicly available encrypted DNS providers.

8. SOAR With DNS

For some security practitioners, security orchestration and automated response (SOAR) is the holy grail. It automates mundane tasks and reduces response time for faster remediation. In this book, we have provided details on how DNS is a treasure trove of information that you should tap into. DNS can integrate with SIEM and IPAM systems and can help you achieve the goal of quickly detecting threats and providing full or partial automated responses.

Closing Comments

We hope the information and suggestions we have collected here provide you with some near-term actionable items, and the book continues to be a general reference guide for you in the future. We also hope you are better equipped now to combat DNS insecurity, and are ready to unleash the potential of DNS as part of your security strategy.

This Book at a Glance

This section is for people who like to skip to the end of the movie or the book. We are not judging—we know your time is valuable, so we tried our best to condense the entire book to a couple of pages.

In Chapter 1, we provide an overview of the structure of the book along with a quick primer on DNS. We learn the basic components of DNS communication — stub resolver (client), recursive resolver (seeker of answers), authoritative servers (keeper of answers), and a quick example of what a resource record looks like.

In Chapter 2, "DNS and Malware," we look at how threat actors use DNS in various stages of the Cyber Kill Chain model. The two case studies, WannaCry and Black KingDom, show how malware uses DNS to evade early detection. We learn that blocking by IP address alone isn't sufficient and blocking by domain lookup is not only more efficient but also necessary.

In Chapter 3, "Look-Alike Domains," we delve into several ways scammers can fool victims by using text characters that look similar, such as the early case of PayPai to scam users of PayPal. Newer Internationalized Domain Names (IDNs) and Top-Level Domains (TLDs) make detection even harder, with not only more character sets, such as the English o and Greek ο (omicron), but also different policies for each TLD that make it possible for threat actors to acquire legitimate-looking domain names, such as `netflix.soy.`, among others. We introduced the filtering mechanism of Response Policy Zone (RPZ), which acts on threat intelligence data to block, drop, rewrite, or allow DNS responses.

In Chapter 4, "Domain Generation Algorithms (DGAs)," we see how threat actors hide their tracks with numerous domain generation algorithm names, ranging from dozens of names per day generated by Zloader, to tens of thousands of names used by Conficker. These malicious names add up to millions, making them difficult to block with traditional blocklists. A "guilty until proven innocent" approach is introduced, in the form of Newly Registered Domains (NRD) or Newly Observed Domains (NOD). Because DGA domain names rarely overlap, the silver lining for security professionals is that we can usually tell what malware is active simply by monitoring what domain names are queried.

In Chapter 5, "DNS Tunneling," we explore how arbitrary data can be sent and received over DNS. However, due to protocol limitations, attackers must break messages down to small chunks and hide them in the most commonly used DNS resource record types, such as A, AAAA, CNAME, and TXT. The case study of InvisiMole illustrates that through the forwarding nature of DNS communications, attackers can control infected devices even without direct Internet access.

In Chapter 6, "Data Exfiltration," we see how attackers could build a one-way tunnel and siphon data out of a network over DNS. The case study of AlinaPOS shows the combination of techniques of look-alike domains and data exfiltration. We also discuss how these queries have specific patterns that we can detect through deep query inspection by looking at length and entropy of labels, frequency and number of total queries, and lexical analysis. We also note how quickly the attacks have evolved in the case study SUNBURST, and how malicious actors are already accounting for some of the detection techniques, such as domain name reputation.

In Chapter 7, "Cache Poisoning and DNSSEC," we discuss the ongoing threat of cache poisoning, which is a protocol-level vulnerability that can only be completely fixed by implementing DNSSEC signing and validation. We observe how attackers play on the rules of probability to poison the DNS cache, and the two case studies show the original Kaminsky vulnerability, as well as the recent SadDNS techniques. We end the chapter by urging readers to deploy DNSSEC, at the very least on their recursive resolvers and on their authoritative servers if possible. We also recommend the use of DNS cookies as an interim security measure.

In Chapter 8, "Encrypted DNS," we explore the issue with privacy in the DNS last-mile communication. We cover the two leading standards, DNS over TLS (DoT) and DNS over HTTPS (DoH). We point out that publicly available encrypted DNS servers present an issue for the enterprise, because not only could they break internal DNS resolution but users would also bypass security policy and protections. The case study of GodLUA illustrates that threat actors were quick to the scene, leveraging DoH to hide their tracks. We wind up the chapter by urging readers to provide encrypted DNS services internally at the enterprise level and to block access to public DoT and DoH servers.

In Chapter 9, "DNS Attacks Against Clients," we look at threats against client DNS behaviors such as HOSTS file tampering, cache manipulation, and redirection to unauthorized DNS servers. We list types of DNS clients to be aware of: classic clients, mobile clients, IoT devices, and web browsers.

In Chapter 10, "Domain Hijacking," we expand the general term "hijacking" into three categories: infrastructure tampering; expired and dormant domains; and record abandonment. For each category, we provide a corresponding case study, showing that even companies with the best cyber security defenses can be compromised through something as fundamental as domain name registration or failing to remove unused DNS records.

In Chapter 11, "DNS and Zero Trust Architecture," we discuss the role DDI can play as a foundational layer in a Zero Trust environment. Security professionals can leverage DNS to detect and mitigate threats such as data exfiltration. Furthermore, advanced environments can use DNS to build and reinforce dynamic security policy, an approach we refer to as D-NAP (DNS and Network Assured Policy).

In Chapter 12, "Conclusion," we summarize the key takeaways from the book and provide eight ways security organizations can harden their defenses against DNS-based threats.

About the Author – Joshua M. Kuo

Joshua M. Kuo dreamed of becoming a novelist growing up in Taiwan but ended up studying computer science in Hawaii. He has been working in the technology field since the late 1990s, wearing many different hats in areas of programming, system administration, network architecture, information security, technical training, and consulting. Josh is passionate about sharing what he knows with others around him, sometimes a little too much. When not preaching about DNS and technology as the Senior Educator at Infoblox, Josh enjoys spending time with his young children, cooking in his kitchen and backyard, and reading about obscure topics in history (such as the history of circus, cod, and the measurement system). Josh currently lives in North Carolina with his family.

About the Author – Ross Gibson, J.D.

Ross Gibson, J.D. is currently a Principal Solutions Architect / Global SME for Infoblox, Inc., where he focuses on DNS security, DNS architecture, global server load balancing (GSLB), DHCP architecture, and IP address management. He holds a B.S. in Commerce from the University of Virginia, and a Juris Doctorate from the University of Richmond School of Law. He brings more than 20 years of experience in the networking industry along with his legal training and experience to his current role. Ross and his wife, Jenn, and their sons Pierce and Evan live in Richmond, Virginia with their Labrador, Aspen. When he and Jenn aren't busy cheering on their boys at baseball games and concerts, he enjoys playing and recording music as well as cooking barbecue and pizza.